AF883538

Springer Theses

Recognizing Outstanding Ph.D. Research

Aims and Scope

The series "Springer Theses" brings together a selection of the very best Ph.D. theses from around the world and across the physical sciences. Each volume has been selected for its scientific excellence by an internationally renowned institute or department. For greater accessibility to non-specialists, the published versions include an extended introduction, as well as a foreword by the student's supervisor explaining the special relevance of the work for the field. The series as a whole provides a valuable resource both for newcomers to the research fields, and for other scientists seeking detailed background information on special questions. Finally, it serves as an accredited documentation of the valuable contributions made by today's younger generation of scientists.

How to Apply

Theses prepared at internationally leading universities and institutes may be nominated for publication by supervisors with the endorsement of the head of department. Each such department is entitled to nominate one thesis per year. Enquiries and requests for a nomination form should be sent, together with the thesis, to Ms. Nathalie Berg (see link on right).

Eligibility Criteria

To qualify for consideration, theses should fulfil the following criteria:

- They must be written in good English.
- The topic should fall within the confines of Chemistry, Physics, Earth Sciences, Engineering and related interdisciplinary fields such as Materials, Nanoscience, Chemical Engineering, Complex Systems and Biophysics.
- The work reported in the thesis must represent a significant scientific advance.
- A maximum of 30% of the text may be a verbatim reproduction from the author's previous publications.
- They must have been examined and passed during the 12 months prior to nomination.
- The theses should have a clearly defined structure including an introduction accessible to new PhD students and scientists not expert in the relevant field.
- After acceptance, theses should be supplemented by a "Supervisor's Foreword" outlining the wider scientific relevance of the work.

Indexed by zbMATH.

Kangjia Fu

Design and Prediction of In-orbit Dynamic Behavior of Deployable Mesh Antennas

Springer

Kangjin Fu
Tsinghua University
Beijing, Beijing, China

ISSN 2190-5053 ISSN 2190-5061 (electronic)
Springer Theses
ISBN 978-981-92-0139-6 ISBN 978-981-92-0140-2 (eBook)
https://doi.org/10.1007/978-981-92-0140-2

Jointly published with Tsinghua University Press

The print edition is not for sale in China (Mainland). Customers from China (Mainland) please order the print book from: Tsinghua University Press.

© Tsinghua University Press 2026

This work is subject to copyright. All rights are solely and exclusively licensed by the Publisher, whether the whole or part of the material is concerned, specifically the rights of reprinting, reuse of illustrations, recitation, broadcasting, reproduction on microfilms or in any other physical way, and transmission or information storage and retrieval, electronic adaptation, computer software, or by similar or dissimilar methodology now known or hereafter developed.
The use of general descriptive names, registered names, trademarks, service marks, etc. in this publication does not imply, even in the absence of a specific statement, that such names are exempt from the relevant protective laws and regulations and therefore free for general use.
The publishers, the authors, and the editors are safe to assume that the advice and information in this book are believed to be true and accurate at the date of publication. Neither the publishers nor the authors or the editors give a warranty, express or implied, with respect to the material contained herein or for any errors or omissions that may have been made. The publishers remain neutral with regard to jurisdictional claims in published maps and institutional affiliations.

This Springer imprint is published by the registered company Springer Nature Singapore Pte Ltd.
The registered company address is: 152 Beach Road, #21-01/04 Gateway East, Singapore 189721, Singapore

If disposing of this product, please recycle the paper.

Foreword

Large deployable mesh antennas are among the most valuable appendages of satellites; they are used for certain sophisticated tasks, such as high-resolution Earth observations and instant telecommunications. As a core technology in the fields of communication and national defense, it has been listed in the "Global Engineering Frontiers 2020" issued by the Chinese Academy of Engineering as one of the top ten engineering research frontiers in the fields of machinery and delivery. Structurally, this kind of mesh antennas is primarily composed of two components, a foldable supporting truss and a flexible cable network. Each of these two components fulfills a central function. The truss is in charge of unfolding the antenna from the launch-ready, stowed configuration to the working configuration in orbit. Once unfolded, the cable network is in charge of forming a geometric paraboloidal shape to reflect signals via the attached metallic mesh. If the truss fails to unfold the entire system, or the cable network fails to form the desired geometric shape within enough accuracy, the antennas cannot work properly.

Research on guaranteeing the success of antennas has boosted the developments of several challenging interdisciplinary mechanical problems. I list a few of them:

1. Modeling, simulation, and design of complex flexible mechanisms. The deployable truss is a typical flexible mechanism suffering large spacial displacement and rotation; meanwhile, it is unfolded by a cable-pulley mechanism and active motors. Understanding and predicting deployment behaviors of the entire system, such as the loads on flexible components and driving forces of the motors, is of great concern. A flexible multibody system simulation is supposed to be a promising way to study the deployment dynamics. However, building a full-scale model to simulate the deployment dynamics is an extremely difficult task. It involves a lot of aspects on studying flexible multibody system and has attracted great research attention in the past years. For example, using Lie groups to formulate 3-dimensional movement of flexible beams and shells. Formulating flexible beam with warping effects. Modeling contacting problems between cable and pulleys with arbitrary Lagrangian-Euler formulations.

2. Designing and evaluation of gravity compensation systems for flexible mechanisms to predict the in-obit behavior. Deploying mesh antennas in-orbit is a delicate and precise process. Thus, on-ground experiments have been conducted to attempt to predict the in-orbit deployment performance of mesh antennas. However, gravitational field on Earth presents a great obstacle in achieving this goal. It is necessary to understand and evaluate how effectively a gravity compensation system can counteract effects of gravity. Nowadays, research is actively focused on developing new gravity compensation mechanisms, either passive or active, to fulfill the on-ground verification purpose.
3. Design and adjusting a flexible cable network structure to form a large paraboloidal shape with enough accuracy. This is a tough job for several reasons. Firstly, the cables may lose tension if the tensile forces are not loaded property, then the network can generate rigid body motion and then result in large shape errors. Secondly, the cable network contains a large number of cable segments, and each of them has a manufacturing errors. These errors also contribute to shape errors. Thirdly, the flexible deformation of cable networks under tensile forces is another source of shape error. Therefore, to make the shape error controllable, it is demanded to understand the governing law adjusting shape of flexible cable network.

Designing a mesh antenna has to deal with the above-mentioned several problems. One of the core dynamic problems involved in its research is to ensure the successful unfolding and the formation of a high precision configuration of the antenna. Dr. Kangjia Fu has studied them when designing a type of mesh antenna, which is known as AstroMesh-II. He proposed a new modeling method to make the simulation a full-scale AstroMesh-II feasible. With the help of this model, he designed a synchronization mechanism to reduce the loads on flexible components and verified the result with an experiment. Then, he studied the design law of gravity compensation system combining theoretically analysis, numerical simulations, and experimental verifications. Finally, he tried to understand how to keep all of the cable network taut under tensile forces. I believe Dr. Fu's thesis is a great theoretical guide for designing mesh antennas. The research results can provide theoretical and design references for large deployable antennas and can also be extended to other space flexible structures such as solar sails and tensegrities. It is of great significance to enrich and develop the in-orbit dynamic prediction and the on-ground experimental design of space flexible structures.

Beijing, China
November 2021

Zhihua Zhao

Related Publications in This Thesis

[1] **Kangjia Fu**, Zhihua Zhao*, Gexue Ren, Yong Xiao, Tao Feng, Jungang Yang, Paolo Gasbarri. From multiscale modeling to design of synchronization mechanisms in mesh antennas[J]. Acta Astronautica, 2019, 159: 156-165.

[2] **Kangjia Fu**, Jianbin Du, Jinyou Li, Zhihua Zhao*. Robust design of tension truss antennas against variation in tension forces[J]. AIAA Journal, 2018, 55: 1450-1459.

[3] Zhihua Zhao, **Kangjia Fu***, Meng Li, Jinyou Li, Yong Xiao. Gravity compensation system of mesh antennas for in-orbit prediction of deployment dynamics[J]. Acta Astronautica, 2020, 167: 1-13.

Preface

Deployable mesh antennas are central units of high-orbit satellites to gather electromagnetic signals. They are composed of flexible cable networks attached with metal mesh to reflect signals and deployable trusses to unfold and support the cable networks for forming a parabolic surface. The circumferential periodic trusses represent an advanced technique for deployable antennas due to their large stiffness-to-mass ratio and high package efficiency. The main tasks in designing mesh antennas are to ensure a successful in-orbit deployment and to form a paraboloidal reflector with sufficiently high shape precision. However, it is difficult for on-ground experiments to predict the in-orbit deployment behavior and the surface accuracy after deployment due to the gravitational difference between on-ground and in-orbit environment. Therefore, it is necessary to use simulation to optimize the designs of the deployable truss, cable networks, as well as gravity compensation systems, to minimize the difference between the in-orbit and on-ground behavior of mesh antennas, and to achieve an accurate in-orbit prediction. This thesis addresses the following three aspects:

A multiscale dynamic method for modeling cable-pulley systems is proposed. In this method, the cable is separated into two segments from the contact border between the pulley and the cable. The non-contact segment is modeled with variable-length cable elements based on the Arbitrary Lagrangian-Eulerian formulation, and the contact segment is merely treated as a virtual segment to dynamically locate the contact border. This method avoids the problems of dense mesh and small step caused by traditional contact modeling methods, greatly improves the computational efficiency, and makes it possible to simulate antenna deployment with multiscale nature of time and length. Furthermore, two kinds of synchronization mechanisms with variant-wrap-angle and constant-wrap-angle were designed to synchronize each bay of the deployable trusses. The proposed multiscale method was used to evaluate the effects of synchronization mechanisms on the deployment dynamics of mesh antennas. Simulation results show that the variant-wrap-angle design leads to remarkable asynchronous deployment with a sharp increase of driving forces by motors, while the constant-wrap-angle design leads to a nearly synchronous deployment accompanied by remarkably small driving forces.

A design methodology for achieving a robust cable network against variation in tension forces was proposed, which reduces the accuracy requirements of the tension forces to guarantee the tautness of the cable network. The method consists of two parts: first, a force tolerance was defined to quantitatively characterize the robustness of the cable network to tension forces; second, an optimization algorithm was established to enlarge the force tolerance by optimizing the tension forces and the geometry of the cable network. The proposed methodology was applied to design the mesh for a 6-bay front-fed antenna and a 30-bay offset antenna. Results show that after optimization, the force tolerances were enlarged up to about four and two times without notably altering the tessellation error, which ensures the tautness of the cable network with or without gravity. The robust design provides a possible technical path to ensure the shape consistency of the cable network between on-ground and in-orbit environment.

The effects of on-ground gravity compensation systems for webs on deployment dynamics of mesh antennas were revealed. In terms of constant-length offloading and constant-force offloading, the position and number of suspension nodes were evaluated on the performance of gravity compensation. Simulation results show that without offloading for webs, the gravitational energy of the webs increases quickly during the early unfolding period since the center of mass of the webs goes up, resulting in larger driving forces for the on-ground test to unfold the mesh antenna. When the webs are offloaded, it proves that too few or too many suspension nodes will add more constraints to the deployment, and a moderate number of suspension nodes with nearly equal distribution on the webs is a better choice to reduce the difference between on-ground and in-orbit deployment.

Beijing, China

Kangjia Fu

Acknowledgment

I would like to extend my sincere gratitude to my supervisor, Associate Professor Zhihua Zhao. At the beginning of my PhD journey, when I often felt uncertain, Professor Zhao provided patient and meticulous guidance. He helped me master fundamental research skills, fostered my enthusiasm for scientific exploration, and supported my transition from an undergraduate to a doctoral researcher. During challenging research periods, his profound knowledge, insightful perspective, and creative thinking greatly broadened my horizons, inspired me to pursue innovation, and allowed me to experience the joy of scientific inquiry. I vividly remember the night before my first academic conference when Professor Zhao stayed up with me until 4 a.m. refining my presentation. For my first academic paper, he reviewed every sentence with utmost care. While studying abroad, whenever I encountered challenges in research or daily life, I often asked myself how Professor Zhao would approach the situation.A lab mate once remarked that I appeared to carry some of Professor Zhao's influence, a comment I regard as a significant compliment. The lessons I have learned from Professor Zhao will benefit me throughout my lifetime.

I am also deeply thankful to Professor Gexue Ren. Whenever I felt inclined to relax, I recalled his consistent habit of working in the lab on weekends and holidays. In the early stages of my doctoral studies, when I struggled to immerse myself in research, my conversations with Professor Ren always provided valuable inspiration. Whenever we met, he kindly inquired about my progress and offered warm encouragement. His diligence and dedication will continue to serve as a model in my future career.

I wish to express my sincere appreciation to Professor Lihua Jin for providing me the opportunity to conduct research on mechanical metamaterials at UCLA. My study abroad experience stands out as one of the most rewarding and memorable chapters of my doctoral journey. I am also deeply grateful to Professor Jianbin Du at Tsinghua University for his invaluable guidance in structural optimization. My heartfelt thanks extend to Yongpeng Gu and Yun Peng for their assistance with multibody dynamics and coding, as well as to Huan Zhang and Jiawei He for their support in various aspects of daily life. Additionally, I would like to thank Meng Li, Jinyou Li, and Jiang Cui for their constructive suggestions regarding my dissertation. Finally, I wish to express my gratitude to every member of our lab for their unwavering support and camaraderie throughout my PhD studies.

I would like to express my deepest gratitude to my parents for their guidance and support throughout my journey from childhood to the completion of my doctorate. I am also profoundly thankful to my brother, with whom I shared a joyful childhood and countless cherished memories as we grew up together. Finally, I extend my heartfelt thanks and love to my wife, whose unwavering support and dedication to our family have enabled me to fully devote myself to my research and studies. I am equally grateful to my parents-in-law for their steadfast support to us. I am also thankful to my son, whose arrival has brought our family boundless joy and happiness.

I extend my sincere gratitude to the professors who reviewed this dissertation for their invaluable suggestions. I also wish to express my appreciation to the Tsinghua University Outstanding Doctoral Dissertation Publication Fund for its support. I am grateful to my editor, Yang Cheng, for his dedicated work, and to all the editors and staff at Tsinghua University Press and Springer for their diligent efforts that made the publication of this thesis possible.

Contents

Abbreviation

A: Area of cross section
ALE: Arbitrary Lagrange-Euler
β: Dummy variable
$\boldsymbol{C}$, C: Constraint equation(s)
d: Wrapping direction of the cable on a pulley
e_{hp}: Half-path-length RMS error
e_{hpbf}: Half-path-length best-fit RMS error
$e_{\mathrm{threshold}}$: Threshold of tessellation error of the web
E_{truss}: Potential of the ring truss
E_{web}: Potential of cable networks
f_{trans}: Friction force of sliding joints
g: Acceleration of gravity
G: Antenna gain
$\boldsymbol{M}$: Mass matrix
M_{bend}: Peak bending moment of the ring truss
$\boldsymbol{N}$: Shape function
P: Center of the pulley
$\boldsymbol{q}$: Generalized coordinate
$\boldsymbol{Q}$: Generalized force
RMS: Root mean square
sign(x): Sign function
$\boldsymbol{t}$: Tangent direction of the pulley
$\boldsymbol{t}^f$: Tension of the cable networks
T_{driving}: Peak driving force by motor
v_{motor}: Reel-in velocity by motor
η: Coefficient of tension decay of the cable passing through pulleys
μ: Coefficient of friction of a sliding joint along the batten
δ: Force tolerance
δ^f: Absolute force tolerance

$\delta^{rel.}$	Relative force tolerance
ρ:	Density
$\boldsymbol{\psi}$:	Cartesian rotation vector
λ:	Vector of the Lagrange multipliers
$\boldsymbol{\varphi}$, φ:	Wrapping angle of the pulley
$\boldsymbol{\Lambda}$:	Coordinate transformation matrix

Chapter 1
Introduction

1.1 Background and Significance

In August 2016, Chinese high-orbit satellite was successfully launched from Xichang Satellite Launch Center, realizing full coverage communication to the domestic and surrounding sea areas, making "not in the service area" no longer happen, and opening a new era of Chinese autonomous satellite telephone. In May 2018, Queqiao, Chang'e-4s relay satellite took the Long March 4C launch vehicle into the Earth-Moon transfer orbit with the perigee at about 200 km and the apogee at about 400,000 km, becoming the first special relay communication satellite outside the earth orbit in human history. The satellite antenna mounted on the Queqiao is designed with an deployable umbrella-like parabolic surface, which is folded and closed during rocket transport to save storage space, and deploys and maintains its parabolic configuration after the satellite is in orbit, with a diameter of 4.2 m after full deployment. This is a new breakthrough in the field of lunar exploration in China, and also marks that China has taken the lead in mastering the earth-moon relay communication technology.

The above-mentioned high-orbit satellite and the Queqiao relay satellite can be used as the epitome of the development of Chinese satellite in the past decade. Looking around the world, space technology represented by satellite antennas has also developed rapidly in recent years, providing strong technical support for communication relay, deep space detection, meteorological observation and other applications [1]. Since the antenna gain is proportional to the square of the aperture [2], in order to capture small power signals and meet the demand of high gain and high coverage, large aperture antenna with feature size of tens of meters [3, 4] was urgently needed. For example, the aperture of Hoop Reflector Antenna proposed by Harris Company has reached 22 m, and the maximum deployed aperture of the starlight cover (Exo-S Shade) proposed by the JPL can reach 34 m [5]. However, the maximum aperture of the payload in current rocket is about 5 m. The contradiction

© The Author(s), under exclusive license to Springer Nature Singapore Pte Ltd. 2026

K. Fu, *Design and Prediction of In-orbit Dynamic Behavior of Deployable Mesh Antennas*, Springer Theses, https://doi.org/10.1007/978-981-92-0140-2_1

between the rocket limitation and the increasing aperture of the satellite antenna makes the antenna possess the characteristics of scalability [6]. That is, the antenna is folded in the rocket while the antenna unfolds in orbit.

Over the past four decades, deployable satellite antennas have been proposed in a variety of ways. According to configurations and deployment ways, they can be divided into three basic types: rigid surface deployable antenna, inflatable antenna and deployable mesh antenna [7]. The rigid surface deployable antenna [8] is composed of some approximate rigid surfaces, and the precision of the surface is quite high, but the structure is bulky and the stowed volume is large. Hence it is difficult for antennas with large aperture. The inflatable antenna [9, 10] belongs to a kind of inflatable film structures with high package efficiency. However, the precision of the surface is poor, which is restricted by the material, process and other factors, resulting in that it still remains in the experimental stage. As shown in Fig. 1.1, a deployable mesh antenna [11–13] is made of flexible mesh, which is light and easy to unfold and capable of combining with the support structure, such as umbrella mesh antenna represented by and folded ribs [14] and a ring mesh antenna supported by the ring truss [15, 16], etc. Mechanically, the umbrella antenna [17, 18] is equivalent to a cantilever beam with radial radiation, while the ring mesh antenna [19] is similar to a closed thin shell. In contrast, the ring mesh antenna has a high specific stiffness and a high package ratio, which is an ideal structure for realizing large aperture antennas.

The mesh antenna Astromesh [19] was proposed by Northrop Grumman Corporation, and it was first successfully applied to communications satellites Thuraya in May 2000, and subsequently to MBSat, Inmarsat 4, Alphasat and recent SMAP projects, with a success rate of 100% in the current public literature and reports. As shown in Fig. 1.2, the mesh antenna is composed of two parts: a

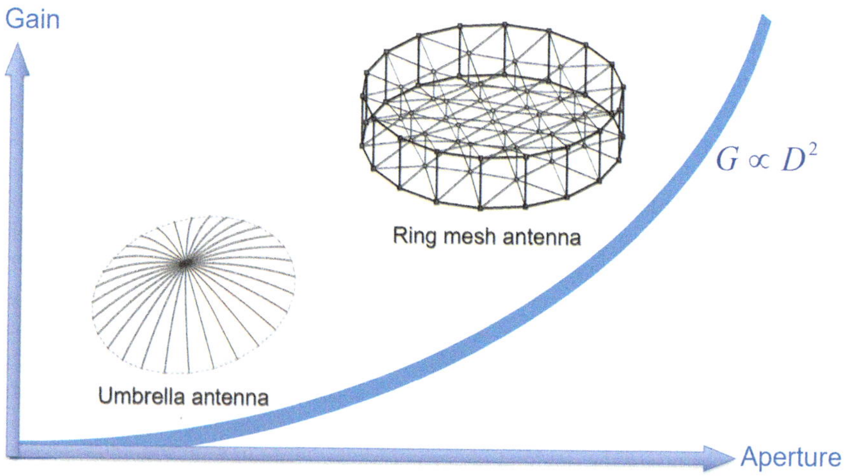

Fig. 1.1 Different types of deployable mesh antennas, where G denotes gain and D signifies aperture

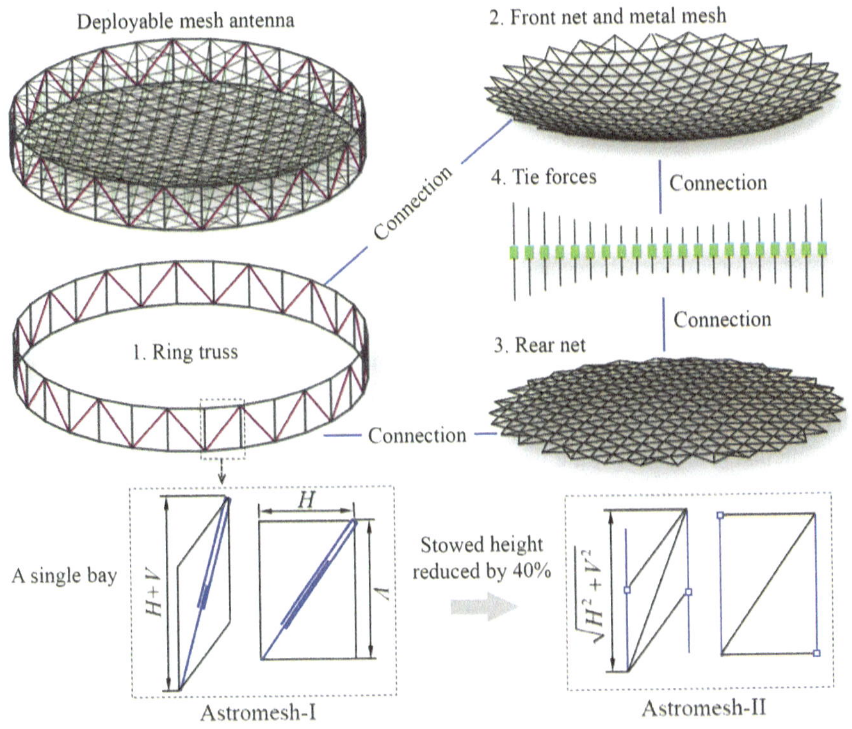

Fig. 1.2 Deployable mesh antennas: Astromesh-I and Astromesh-II

deployable ring truss and cable networks [20]. If the members of the ring truss are considered as rigid parts, the ring truss can be regarded as a deployable mechanism with a single degree of freedom [21, 22] which is unfolded by motors. The cable networks consist of a front net, a rear net, a metal mesh attached to the front net (used to receive and send electromagnetic signals, not shown in the diagram), and a tension tie spring (also known as a tension array) between the front and rear net [23]. When the antenna is fully unfolded, the tie springs will tighten the nodes corresponding to the front net and the rear net, thus maintaining the parabolic configuration and the necessary overall stiffness of the structure.

Figure 1.2 shows two deployment forms of the ring truss, including Astromesh-I [24] and Astromesh-II [25]. Both are composed of similar bays, each of which has two longerons with length H, two battens with length V and a diagonal. For Astromesh-I, each diagonal contains two telescope tubes, and longerons and tubes are hinged with battens. Elongating the diagonals can fold the ring truss, and vice versa. When fully folded, battens are co-linear with longerons, hence the stowed height is $H + V$. The revolute joints of adjacent bays are synchronized by a pair of interlocking gears.

However, Astromesh-II has a different unfolding mechanism, in which each diagonal is in one piece. One end of a longeron hinges with a batten, and the other end connects with the other batten through a translational joint. The translational joints of adjacent bays are synchronized through particularly designed cable-pulley systems. When fully folded, the stowed height is the diagonal length. Among these two models, Astromesh-II has a more complex synchronization mechanism. Nevertheless, owing to the lower stowed height, it is preferred by large mesh antennas for saving storage space. Astromesh-II will become an important choice of large satellite antenna in the future and is also the research object of this thesis.

Whether the antenna can be successfully deployed directly determines the success or failure of the whole satellite mission. Because of the complexity of antenna mechanism and the influence of gravity, it is difficult to give accurate and complete description according to ground experiments. Dynamic simulation provides technical support for antenna design and in-orbit behavior prediction [12, 26–28]. Different from the Astromesh-I [29–32], the Astromesh-II has adopted a new deployment mechanism which lacks a systematic and in-depth understanding. In the antenna design stage, the first problem is to design the proper synchronization mechanism to ensure the synchronous deployment of each bay of the ring truss, and the multiscale characteristics of the synchronization mechanism become the bottleneck restricting the dynamics simulation of antenna. In addition, it is necessary to evaluate the influence of the synchronization mechanism on the antenna deployment dynamics, such as the driving force by motor and the component loads and other parameters. On this basis, it is also necessary to find out the possible problems in unfolding process, find out unreasonable aspects in the current design, and further improve or propose new designs.

As shown in Fig. 1.2, after the antenna is fully deployed, the tie springs make both the front and rear net taut. The metal mesh is attached to the front net and the parabolic surface is fitted with triangular surfaces [33]. The electromagnetic transmitter/receiver located at the focus of the parabolic surface. Compared with rigid surfaces, the flexibility of cable networks leads to a great challenge in realizing parabolic configuration with high precision. It usually requires that the root mean square (RMS) error of the surface can be controlled within the order of millimeters, and that the profile error cannot change with the increase of the diameter of the antenna [34]. As a quantitative example, an S-band mission with a frequency of 4.0 GHz (wavelength 75 mm) requires a shape error as small as 1.5 mm, regardless of the aperture size: for instance, 6 m or 30 m [35, 36]. Therefore, investigations on factors and methods that can affect and/or improve the shape precision are of great concern.

To sum up, there are two core tasks for the deployable mesh antenna: The first is to make the antenna deploy on orbit successfully, which involves determining whether the motor driving force is enough to ensure the antenna is fully deployed, and whether the strength of each component of the antenna is enough to withstand the load during unfolding process. The second is to form a high-precision paraboloid to send and receive electromagnetic signals after the antenna is fully deployed. The core problem is how to achieve a paraboloidal surface with required precision in orbit.

1.2 Current Research and Unresolved Issues

1.2.1 Design of Deployable Structures

Driven by motors, the ring truss realizes the transition from the stowed state to the deployed state. Since the motors are located on a certain bay of the ring truss, in order to ensure that each bay of the ring truss can be deployed, the ring truss should have a single degree of freedom if each component of the ring truss can be regarded as rigid. At this time, it is necessary to introduce a specific design to ensure the synchronization of each bay of the ring truss, which is called the synchronization mechanism. The mesh antennas including Astromesh-I and Astromesh-II adopt different synchronization mechanisms as shown in Fig. 1.3. For Astromesh-I, each diagonal contains a pair of telescope tubes, longerons and battens. Reeling-in the driving cable by motor can shorten the diagonals, resulting in unfolding the ring truss. At the same time, a pair of synchronous gears is introduced between the adjacent two bays to ensure that the rotation angles of the adjacent two bays are equal. An asynchronous unfolding process from the near-motor side to the far-motor side in sequence occurs during deployment. For Astromesh-II, the synchronization mechanism changes from the synchronous gears to synchronization cable-pulley systems, which aim to ensure the same displacement between the two adjacent translational joints along battens. Taking a single bay as an example, there are two translational joints, and the original degree of freedom is two. When the left slider moves down after using the synchronization cable-pulley system, the synchronization cable will pull the right slider up; when the left slider moves up, the synchronization cable will pull the right slider down. Hence the displacement of the two adjacent sliders is mutually restricted. The goal is to make the ring truss into a system with single degree of freedom.

Theoretically, if all components of the ring truss are considered as rigid bodies, the synchronous gears ensure that the ring truss of Astromesh-I has only one degree

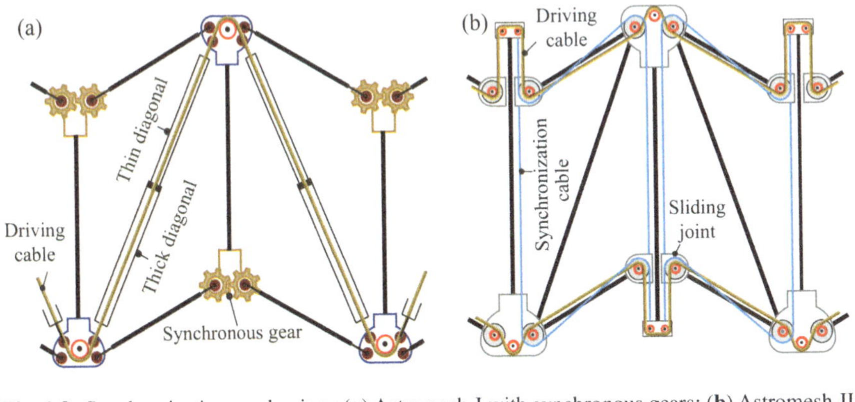

Fig. 1.3 Synchronization mechanism: (**a**) Astromesh-I with synchronous gears; (**b**) Astromesh-II with synchronization cable-pulley system

of freedom, and each bay will be completely synchronized. In fact, the flexibility of the truss and the tension decay of driving cables through pulleys cause an asynchronous deployment. The ground unfolding experiment of Astromesh-I shows that the bays deploy from the near-motor side to the far-motor side in sequence. Specifically, the bays near the motors begin to latch at first, and the bays far from the motors latch at last. Considering that the flexibility of the ring truss becomes remarkable with the increase of antenna diameter, the rigid assumption of the ring truss can no longer correctly describe the dynamic behavior of deployment. Peng Yun [37], Li Pei [38], Liu [39, 40] and others have established the flexible dynamic models of Astromesh-I respectively. The simulation results showed the asynchronization of deployment.

Current research results show that the asynchronous deployment of Astromesh-I [41] has a significant effect on the deployment dynamics, such as the loads of components (the bending moments of the truss mainly), and the driving force of motors. Compared with the synchronous gears in Astromesh-I, the synchronization cable-pulley systems used in Astromesh-II have a more complex design to ensure synchronization. For Astromesh-II, it is necessary to evaluate the effect of synchronization mechanism on deployment dynamics. On this basis, problems and unreasonable designs can be found out, and new designs and improvements can be further proposed.

1.2.2 Modeling and Simulation of Deployable Antennas

Recently, the rapid development of flexible multibody system dynamics [42, 43], especially the proposed flexible element with large deformation, provides a powerful tool for modeling and simulation of deployment dynamics of deployable antennas [44]. For modeling methods, multibody system dynamics can be divided into three categories: the floating coordinate formulation, the absolute node coordinate formulation (ANCF), and the geometric accurate formulation. The floating coordinate method [45] is to establish a floating coordinate frame attached on the flexible body, which decomposes the motion of the flexible body into the implicated movement of the floating frame (large scale of the rigid translational and rotational motion) and the small elastic deformation relative to the floating frame, which is suitable for the system with small deformation. The absolute node coordinate formulation [46, 47] was proposed by Shabana, in which the movement and deformation of the flexible body are expressed by the spatial coordinates and gradient coordinates of the element nodes relative to the inertial coordinate frame. On this basis, finite elements including beam, plate, shell and solid are developed with high precision of displacement description. The geometric accurate formulation was first proposed by Simo in the 1990s [48, 49]. For calculation, the curvature is used to describe the deformation of beams and shells in three-dimensional space, and both the rotation and displacement are interpolated independently. Since the beam and shell are mainly subjected to bending load in most cases, it is more suitable to describe the three-dimensional bending deformation of beams and shells.

Jiali Tang [50] proposed a time-varying flexible beam element based on the absolute node coordinate formulation. Both the absolute coordinate and slope of the node are used to describe the configuration of the element, and the influence of the mass flow at borders is considered. Lei Yu [51] established a unified solution framework of flexible multibody dynamics. It provides a complete set of modeling methods for large rotating and deformed flexible bodies by combining one-dimensional cable/beam element, two-dimensional plate/shell element and three-dimensional solid element with existing multibody system algorithms according to the first class of Lagrange equations. Qiang Tian [52] used the method of absolute node coordinates formulation to study the multibody systems with kinematic joints considering the clearance and friction (including revolute joint and spherical joint), and conducted the sensitivity analysis. Difeng Hong [53] proposed the Eulerian Bernoulli beam element described by arbitrary Lagrangian-Eulerian description, and the sliding joint along one-dimensional flexible medium was established. Zhihua Zhao [54] developed the method of geometric accurate beam and shell elements based on rotational vector description. Pei Li [38] proposed a recursive algorithm based on static condensation and multi-layer region decomposition, and the MPI parallel technique was used to solve the dynamic equations of largescale multibody systems efficiently. Yun Peng [37] proposed a variable length cable element with arbitrary Lagrangian-Eulerian description, the material coordinates were introduced into the generalized coordinates of the cable element. Based on this, a simplified modeling method of the cable-pulley system was proposed. In order to improve the computational efficiency, Li Pei et al. [38] conducted the simulation of the deployment of antenna by using parallel calculation, analyzed the influence of cable pretension on the deployment, and discussed the main reasons that caused the asynchronization of the deployment.

A series of modeling and simulation studies on the Astromesh-I have been carried out by Mengliang Zhao, Fulin Guan, etc. [55, 56], in which the dynamic equation of the deployment of the ring truss was established by the generalized inverse matrix method, the deployment of the antenna was analyzed, and the mechanism of friction in the deployment process was studied. Chaohui Qi [57] studied the influence of the parameters of the pulleys on deployment. The relationship between the deployment speed of the antenna and the reel-in velocity of driving cables by motors was given from the kinematics point of view, and the deployment process of the antenna is controlled by controlling the reel-in velocity of the driving cables by motors. Based on the multibody dynamics method, Tuanjie Li [58] established the dynamic model of Astromesh-I, obtained the change of the driving force during deployment, and studied the effects of torsion stiffness of springs and pretension of webs on deployment. Yun Peng [31] established the flexible multibody dynamic model of Astromesh-I, and simulation results showed an asynchronous deployment, which was validated by the on-ground experiment. On this basis, the influences of the asynchronous deployment on the driving force by motors and the load of the ring truss were systematically analyzed. In order to improve the computational efficiency, Pei Li [38] simulated the whole process of the deployment with parallel calculation, analyzed the influence of pretension of cables on deployment, and discussed the main reasons for the asynchronous deployment of the ring truss.

For the cable-pulley systems in Astromesh-I, if the cable-pulley system is modeled by traditional contact method, it requires the element length to be about one tenth of the pulley size for fitting the pulley curvature, resulting in large amount of generalized coordinates and contact detection, as shown in Fig. 1.4. Therefore, the calculation efficiency is too low to fulfill an unfolding simulation of a mesh antenna. Peng et al. [59] proposed an efficient method, which avoided contact detection based on variable-length cable elements with Arbitrary Lagrangian–Eulerian (ALE) formulation, as shown in Fig. 1.5. However, it actually bypasses rather than solves the multiscale problem by treating pulleys as dimensionless points, which is only suitable when the pulley size does not matter, such as the case in Astromesh-I.

Nevertheless, the pulley size is crucial in the synchronization mechanisms of Astromesh-II, since it affects the synchronization conditions of adjacent bays. Therefore, the multiscale nature of the cable-pulley systems must be exactly modeled, and the effects of synchronization on cable-pulley systems on deployment dynamics can be investigated based on this, which will be one of the contents of this thesis.

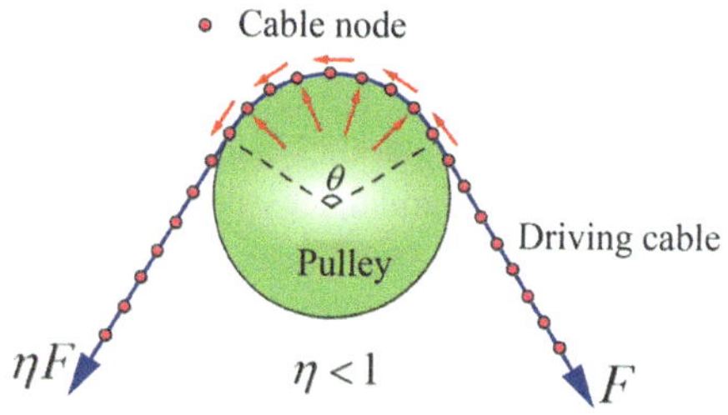

Fig. 1.4 Contact modeling method of the cable-pulley system, where F denotes the cable tension and η signifies the coefficient of tension decay of the cable passing through the pulley

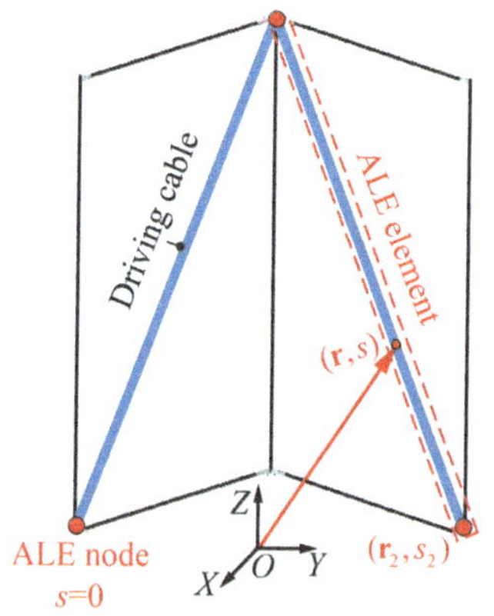

Fig. 1.5 Simplified ALE Modeling Method for Cable-Pulley Systems, where $\boldsymbol{r}$ represents the position coordinate and s denotes the material coordinate

1.2.3 Design of Cable Networks

A mesh antenna comprises a supporting truss and a pair of mirror cable networks mutually tensioned by low-stiffness tie springs. The cable networks consist of a front net, a rear net, a metal mesh attached to the front net, and a tension tie spring (also known as a tension array) between the front and rear net. When the antenna is fully deployed, the tension array will tighten the nodes corresponding to the front net and the rear net, thus maintaining the parabolic configuration and the necessary overall stiffness of the structure. The metal mesh is attached to the front net and the parabolic surface is fitted with triangular surfaces. The electromagnetic transmitter/receiver is located at the focus of the parabolic surface. The quantitative relationship between parabolic gain and shape error [60] as

$$\frac{G}{G_0} = \exp\left[-\left(\frac{4\pi e}{\lambda}\right)^2\right] \tag{1.1}$$

in which G is the gain of the mesh surface, G_0 is the standard parabolic gain, λ is the electromagnetic wave length, and e is the root mean square (RMS) error of the mesh surface. Applications demand surface errors smaller than 1/50th of the wavelength [35, 36]. Therefore, it is a great challenge to achieve a parabolic surface with high precision. The design of cable networks involves two aspects: The first is the design of topology, including the number of network surface nodes and the corresponding connection; The second is the geometric design of cable segments, or the coordinates of cable nodes on the net.

First, at the topological level, Calladine [61, 62] and Tibert [63] proposed that after the mesh antenna is completely unfolded, the whole system including the cable networks and the ring truss should be a statically determinate structure, that is, the degree of freedom of the whole system is exactly 0:

$$0 = 3(n+b) - (b+c+b+3) = 3n + b - c - 3 \tag{1.2}$$

in which b is the number of the bars (the number of bar nodes); n is the number of the cable nodes (the number of tie springs); c is the number of the cables. For the statically determinate design, the shape of the cable network is determined solely by the lengths of the structure members with a given topology. Hence, if the length error is negligible, the cable nodes should be located on the ideal surface. Figure 1.6 shows four webs with different topologies when the bar nodes $b = 30$, in which p represents the number of meshes along the radius. Theoretically, the tessellation error decreases with the increase in the meshes. However, dense mesh leads to the increase in the mass of the cable networks and more errors in the cable lengths. Therefore, a simplified topology should be adopted on the premise of satisfying the required shape precision.

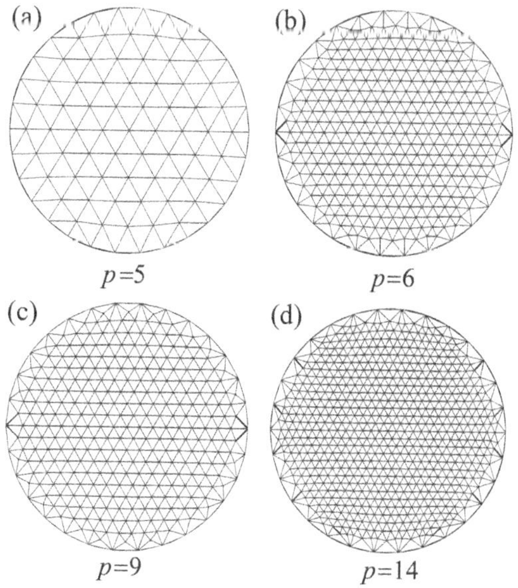

Fig. 1.6 Designs of statically determinate structure with different topologies: the number of meshes increases from (**a**) to (**d**)

Secondly, in the geometrical design of the cable networks, lots of shape finding methods have been proposed in the opening literature [64]. Shek [65] and Linkwitz [66] proposed the force density method, in which the force density is defined as the ratio of cable tension to its length, and the coordinates of each node can be obtained according to the force density of each cable. The advantage of force density method is that the position coordinates of each node can be solved by linear equilibrium equations instead of iteration. Based on the force density method, the minimum length configuration was obtained by Tibert [63, 67] on the premise of equal force density. Recently, the research on the geometric design of cable networks by optimization has been a hot topic. Morterolle et al. [68] proposed the improved force density method to ensure the uniform distribution of the mesh tension. Based on the finite element model of the cable networks, Jiejian Di [69], Dongwu Yang [70], Rui Yang [23], Jingli Du et al. [71] adopted optimization algorithms including genetic algorithm, inverse iteration method, sequential quadratic programming method and minimal norm method to improve the accuracy of the mesh surface. Pei Li [72] considered the influence of flexible deformation of ring truss in the process of shape finding of the cable networks. In addition, studies have been carried out on the influence of length error on shape precision [73–75], and relevant attempts have also been made to adjust the length of several cable segments to improve the shape precision [76, 77] and the electromagnetic performance [78–84].

At present, most studies focus on the geometrical design of the cable networks when the topology is given, and the influences of length error on shape precision are investigated. However, the influence of the tie forces on shape precision cannot be ignored. If the tie forces are not appropriate or there exist large force errors of the tie forces, part of the cable networks may be slack, resulting in remarkable shape error. The relation between the tie forces and mesh surface can be summed up as a forward question and an inverse question. The forward question is how to select the appropriate tie forces to make all the cables taut; The inverse question is how to calculate the inverse problem can be expressed as how to calculate the allowable deviation of the forces that can guarantee tautness of all of the cables.

For the forward question, Mitsugi [85] and Tibert [67] proposed the use of Monte Carlo simulations to check whether cables slackened due to deviations of the tie forces. However, the reverse problem has not yet been investigated, to the best of our knowledge, in the literature. This might be due to the lack of a precise definition of force tolerance and an efficient method to formulate and solve tolerance problems. As a result, no further effort has been made to optimize the cable network to obtain increased force tolerance. Such a large improvement has significant practical implications: tolerance for manufacturing errors can be increased, and the permissible working environment can be expanded.

1.2.4 On-Ground Validation

As an important verification means, on-ground experiments check whether the antenna can be successfully deployed and whether the shape precision of the mesh surface after deployment can meet the requirements [86, 87]. Compared with the in-orbit deployment, on-ground experiments eliminate the cost of launch and have more measurement methods which can monitor various parameters during deployment in real time, such as driving forces by motors and loads of each component. In addition, the kinematic description during deployment can be obtained by using the multi-vision measurement system. However, the primary problem of using on-ground experiments to predict the in-orbit behavior of antennas is the gravity difference. The in-orbit deployment occurs in a microgravity environment while the on-ground deployment occurs under gravity with the gravitational acceleration $g = 9.8$ m/s^2. In order to make the on-ground experiment predict the in-orbit behavior as accurately as possible, namely to achieve the equivalence between in-orbit and on-ground environment, it is necessary to design the corresponding gravity compensation system [88]. It should be mentioned that gravity compensation systems are also called offloading systems in this thesis.

At present, gravity compensation methods can be classified into the following types: free-fall method in falling wells or falling towers, air suspension method, water floating method, and mechanical suspension method [89, 90]. In contrast, the mechanical suspension method is simple, economical, and suitable for gravity compensation of large space deployable structures. At present, the mechanical suspension method is widely used for on-ground experiments such as manipulators [91–97], ground deployment of solar deployable panels [98, 99], and deployable antennas including planar deployable antennas [100], deployable truss antennas [101–103], umbrella-shaped deployable antennas [104], modular truss antennas [86] and deployable mesh antennas [105, 106], etc.

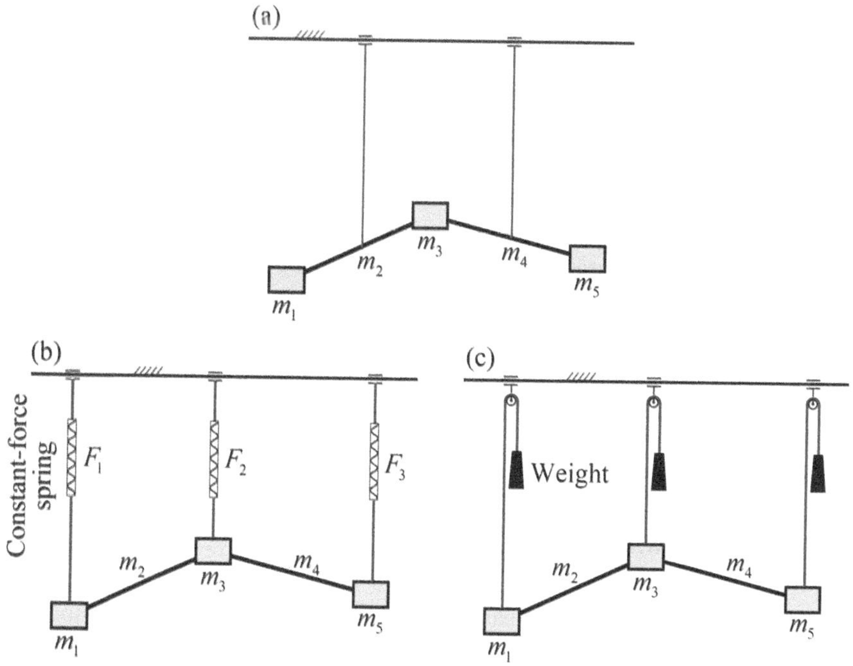

Fig. 1.7 Mechanical suspension method for gravity for gravity compensation: (**a**) Constant-length suspension; (**b**) Constant-force suspension with constant-force springs; (**c**) Constant-force suspension with cable-pulley-weight systems. Here, m1 through m5 correspond to the masses of the components, while F1, F2, and F3 correspond to the forces exerted by the springs

The basic idea of mechanical suspension method to achieve gravity compensation is to provide vertical upward tensions to compensate gravity. The suspension system is mainly composed of two parts including a horizontal follow-up system and vertical suspension strings including constant-length strings and constant-force strings, as shown in Fig. 1.7. As shown in Fig. 1.7a, constant-length strings usually adopt cables with large elastic modulus such as steel wires to ensure that the vertical height of the suspension node remains a constant. By selecting a suitable suspension node, no work is done by gravity as

$$W_G = 0 \tag{1.3}$$

The constant-force suspension is usually realized by constant-force springs or cable-pulley-weight systems, as shown in Fig. 1.7b and c. An appropriate constant force should be selected to ensure that the work done by gravity W_G is completely compensated by the work done by the suspension system W_F, i.e.

$$W_G + W_F = 0 \tag{1.4}$$

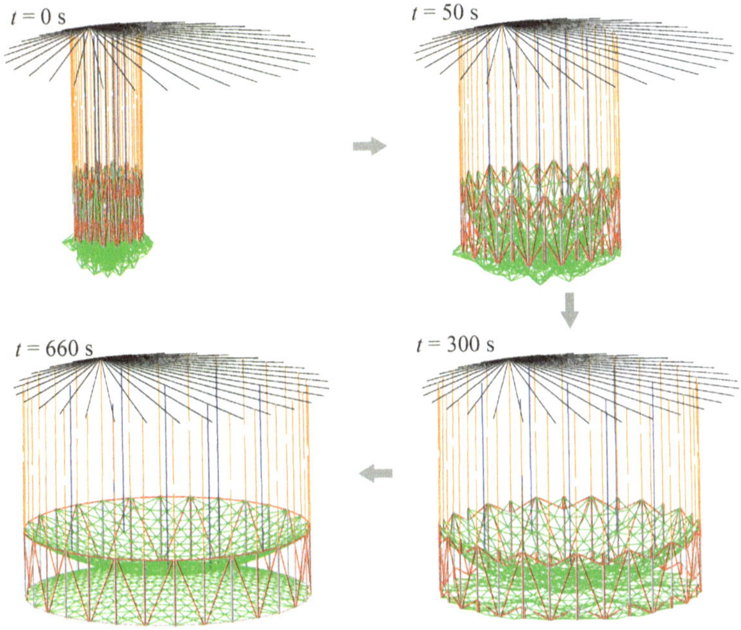

Fig. 1.8 Gravity compensation of the deployable mesh antenna

For deployable mesh antennas, it has remained impossible to eliminate all of the effects of gravity considering the large movement and deformation, as shown in Fig. 1.8. Therefore, there is always some degree of difference between on-ground verification data and in-orbit flight data. These differences should be reduced to the extent possible by designing gravity compensation systems based on a proper and thorough understanding of the dynamical differences between on-ground deployment and in-orbit deployment.

1.3 Research Content of this Thesis

The main tasks in designing mesh antennas are to ensure a successful in-orbit deployment and to form a paraboloidal reflector with sufficient high shape precision. However, it is difficult for on-ground experiments to predict the in-orbit deployment behavior and the surface accuracy after deployment due to the gravitational difference between on-ground and in-orbit environments. Therefore, it is necessary to use simulation to optimize the designs of the deployable truss, cable networks, as well as gravity compensation systems, to minimize the difference between the in-orbit and on-ground behavior of mesh antennas, and to achieve an accurate in-orbit prediction. The main framework of this thesis is shown in Fig. 1.9.

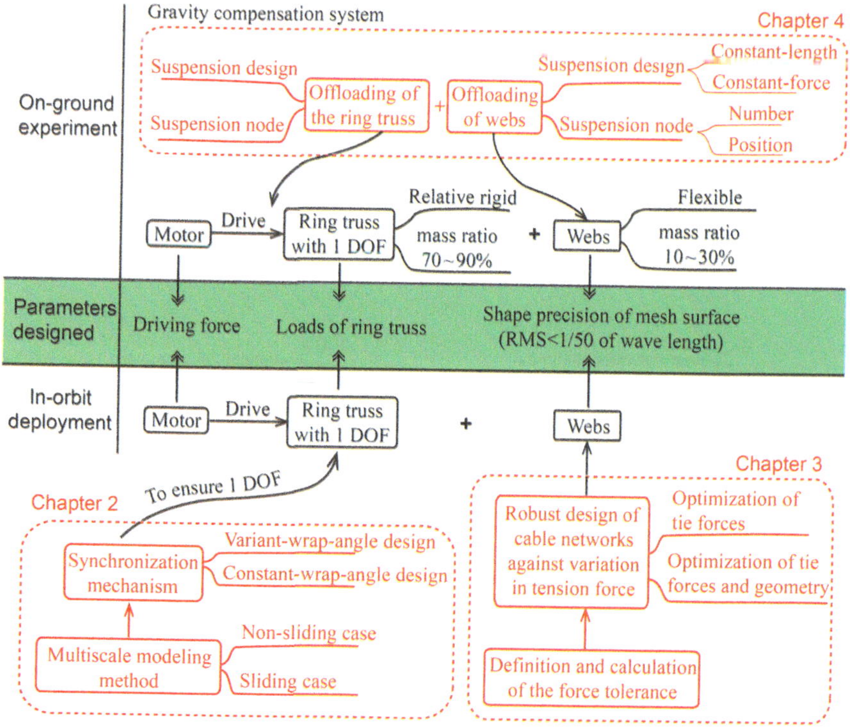

Fig. 1.9 The framework of this thesis

In the second chapter, a multiscale dynamic method for modeling cable-pulley systems was proposed. In this method, the cable is separated into two segments from the contact border between the pulley and the cable. The non-contact segment was modeled with variant-length cable elements based on Arbitrary Lagrangian-Eulerian formulation, and the contact segment was merely treated as a virtual segment to dynamically locate the contact border. This method avoids the problems of dense mesh and small step caused by traditional contact modeling methods, greatly improves the computational efficiency, and makes it possible to simulate antenna deployment with multiscale nature of time and length. Furthermore, two kinds of synchronization mechanisms with variant-wrap-angle and constant-wrap-angle were designed to synchronize each bay of the deployable trusses. The proposed multiscale method was used to evaluate the effects of synchronization mechanisms on the deployment dynamics of mesh antennas. Simulation results show that the variant-wrap-angle design leads to remarkable asynchronous deployment with a sharp increase in driving forces by motors, while the constant-wrap-angle design leads to a nearly synchronous deployment accompanied by remarkably small driving forces.

In the third chapter, a design methodology for achieving a robust cable network against variation in tension forces was proposed, which reduces the accuracy requirement of the tension forces to guarantee the tautness of the cable network. The method consists of two parts: first, a force tolerance was defined to quantitatively characterize the robustness of the cable network to tension forces; second, an optimization algorithm was established to enlarge the force tolerance by optimizing the tension forces and the geometry of the cable network. The proposed methodology was applied to design mesh for a 6-bay front-fed antenna and a 30-bay offset antenna. Results show that after optimization, the force tolerances were enlarged up to about four and two times without notably altering the tessellation error, respectively, which ensures the tautness of the cable network with or without gravity. The robust design provides a possible technical path to ensure the shape consistency of the cable network between on-ground and in-orbit environments.

In the fourth chapter, the effects of on-ground gravity compensation systems for webs on deployment dynamics of mesh antennas were revealed. In terms of the constant-length offloading and constant-force offloading, the position and number of suspension nodes were evaluated on the performance of gravity compensation. Simulation results show that without offloading for webs, gravitational energy of webs increases quickly during the early unfolding period since the mass center of the webs goes up, resulting in larger driving forces for the on-ground test to unfold the mesh antenna. When the webs are offloaded, it proves that too few or too many suspension nodes will add more constraints to the deployment, and a moderate number of suspension nodes with nearly equal distribution on webs is a better choice to reduce the difference between on-ground and in-orbit deployment.

In the fifth chapter, there is a summary and prospect of this dissertation.

Chapter 2
From Modeling to Design of Synchronization Mechanisms in Mesh Antennas

2.1 Introduction

As shown in Fig. 2.1a, the deployable mesh antenna consists of a deployable ring truss and cable networks. If each bar of the ring truss is regarded as a rigid body, the ring truss can be regarded as a deployable mechanism with a single degree of freedom, which is driven by motors. The cable networks contain a front net, a rear net, and a metal mesh attached to the front net (used to reflect the electromagnetic signal, not drawn in the figure), and the tie springs between the front net and the rear net.

Figure 2.1 gives two deployment forms of the ring truss, including Astromesh-I [24] and Astromesh-II [25]. Both are composed of similar bays, each of which has two longerons with length H, two battens with length V and a diagonal. For Astromesh-I, each diagonal contains two telescope tubes, and longerons and tubes are hinged with battens. Elongating the diagonals can fold the ring truss, and vice versa. When fully folded, battens are co-linear with longerons, hence the stowed height is $H + V$. The revolute joints of adjacent bays are synchronized by a pair of interlocking gears. However, Astromesh-II has a different unfolding mechanism, in which each diagonal is in one piece. One end of a longeron hinges with a batten, and the other end connects with the other batten through a translational joint. The translational joints of adjacent bays are synchronized through particularly designed cable-pulley systems. When fully folded, the stowed height is the diagonal length $\sqrt{H^2 + V^2}$. Among these two models, Astromesh-II has a more complex synchronization mechanism. Nevertheless, owing to the lower stowed height, it is preferred by large mesh antennas for saving storage space. Because of the saving of rocket storage height, the mesh antenna Astromesh-II is suitable for designing antenna with larger aperture.

© The Author(s), under exclusive license to Springer Nature Singapore Pte Ltd. 2026

K. Fu, *Design and Prediction of In-orbit Dynamic Behavior of Deployable Mesh Antennas*, Springer Theses, https://doi.org/10.1007/978-981-92-0140-2_2

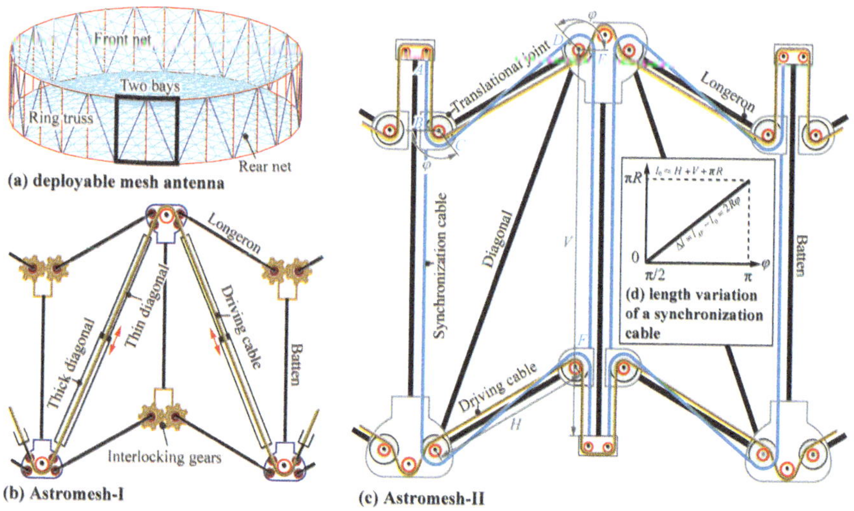

Fig. 2.1 Multiscale problem of deployable mesh antenna: (**a**) deployable mesh antenna; (**b**) Astromesh-I; (**c**) Astromesh-II. Adapted with permission from [107]

2.2 Multiscale Modeling Problems of Mesh Antennas

Previous analysis on Astromesh-I has revealed that the bays unfold with different speeds due to notable flexibility and friction during deployment; and this asynchronization behavior has a substantial effect on the bending moments of the truss and the driving forces by motors. Similarly, it is expected that the synchronization mechanisms of Astromesh-II play an important role in the deployment dynamics. One design of the cable-pulley mechanism is shown in Fig. 2.1c, in which one single synchronization cable passes through two identical pulleys with radius R. There are totally four identical synchronization cables in Fig. 2.1c, illustrated by the blue solid curves. When the bays transform from the deployed to the stowed configuration, geometric analysis shows that the wrap angle at which the cable winds on a pulley, φ, varies roughly from $\pi/2$ to π, assuming the two translational joints move with the same distance; as a consequence, the cable is extended by a magnitude of πR, as shown in Fig. 2.1d. However, the cable is nearly inextensible, so that the adjacent two translational joints cannot be perfectly synchronized. For the design purpose of the synchronization mechanism, it is necessary to evaluate its role on the overall deployment of Astromesh-II and to improve the design.

A flexible multibody simulation is supposed to be a promising way to study the deployment dynamics [10], especially encouraged by the great progress achieved on modeling individual flexible bodies [11–13], and simulating the entire Astromesh-I [8, 15–18]. However, simulating Astromesh-II remains an extremely challenging task. In our opinion, the bottleneck is modeling the synchronization cable-pulley system, stemming from the inherent multiscale problems in terms of both length and time. On length scale, the pulleys are ~10 mm, while the cables are ~10 m, spanning three orders of magnitude; on time scale, the characteristic time of contact event is ~10 ms, while the entire unfolding process lasts for ~1000s, spanning five orders of magnitude. If the cable-pulley system is modeled by traditional contact method, the calculation efficiency is too low to fulfill an unfolding simulation of a mesh antenna. This is mainly caused by two reasons. One is that it requires the element length is about one tenth of the pulley size for fitting the pulley curvature, resulting in large amount of generalized coordinates and contact detection. The other is that the high contact stiffness demands small integration time step. Recently, Peng et al. [31] proposed an efficient method, which avoided contact detection based on variable-length cable elements with Arbitrary Lagrangian–Eulerian (ALE) formulation. However, it actually bypasses rather than solves the multiscale problem by treating pulleys as dimensionless points, which is only suitable when the pulley size does not matter such as the case in Astromesh-I.

Nevertheless, the pulley size is crucial in the synchronization mechanisms of Astromesh-II, since it affects the synchronization conditions of adjacent bays. Therefore, the multiscale nature of the cablepulley systems must be exactly modeled. In this study, we further extended the method proposed by Peng et al., and developed a novel multiscale modeling methodology to preserve pulley radius which takes account of all the relevant multiscale effects. Then, we designed two different synchronization mechanisms and applied the proposed modeling method to systematically studying their corresponding deployment dynamics. Results reveal deep insights on the deployment, such as synchronization performance and the accompanying so-called "butterfly effect" on the driving force. The obtained insights have high potential to promote the actual design of mesh antennas, and the proposed multiscale method can be applied to simulating other similar mechanical systems.

2.3 Multiscale Modeling Methods of Cable-Pulley Systems

The aim of the present work is to develop a novel and efficient modeling methodology to handle the inherent multiscale problem of a cable-pulley system. The basic idea is to formulate the cable-pulley system with a few generalized coordinates while avoiding the need to detect contact events. This is achieved by modeling the non-contact and contact segments separately. As shown in Fig. 2.2, the non-contact part of the cable is meshed using variable-length cable elements based on the ALE formulation [8, 21], illustrated by the solid lines; the contact part is considered as a virtual segment whose arc length is time variant, as illustrated by the dashed line. In this method, the position of the border node between the non-contact segment and contact segment is dynamically located. This approach avoids the contact detection, and as a result, a rough cable mesh is sufficient for accurately modeling the cable-pulley system.

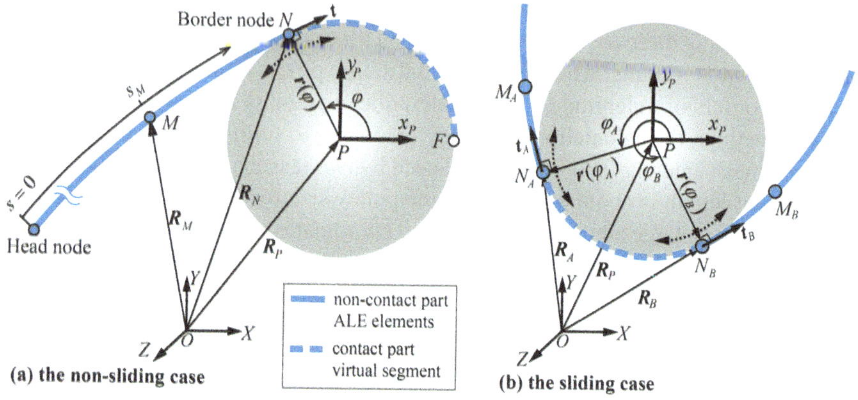

Fig. 2.2 Multiscale modeling of the cable-pulley system: (**a**) the non-sliding case; (**b**) the sliding case. Adapted with permission from [107]

For convenience of formulating the positions of the border nodes, the cable-pulley systems are classified into two categories, as shown in Fig. 2.2. One is the no-sliding case, in which the cable wraps around the pulley with one end F fixed on the pulley, so that the cable does not slide relative to the pulley; there is one border node N. The other is the sliding case, in which the cable passes over and may slide on the pulley; in this case there are two border nodes N_A and N_B.

2.3.1 *Modeling of the Non-contact Segment*

The non-contact cable segment was formulated based on the ALE strategy. The length of an ALE element may vary during wrapping or movement of the cable, and its control volume was chosen as the moving space between the border node N and a material node M, as shown in Fig. 2.2. The non-contact cable part was formulated by ALE cable elements assuming the two end nodes are given. For an ALE element, the generalized coordinates of an ALE element include both global positions and material coordinates:

$$\boldsymbol{q}_{\mathrm{e}} = \left[\boldsymbol{R}_M^{\mathrm{T}} \ \boldsymbol{R}_N^{\mathrm{T}} \ s_M \ s_N\right]^{\mathrm{T}} \tag{2.1}$$

The position $\boldsymbol{R}(s,t) = \boldsymbol{N}_R(s)\boldsymbol{q}_R(t)$ of a point within the ALE cable element is linearly interpolated as

$$\begin{aligned} &\boldsymbol{R}(s,t) = \boldsymbol{N}_R(s)\boldsymbol{q}_R(t), \\ &\boldsymbol{q}_R = \left[\boldsymbol{R}_M^{\mathrm{T}} \quad \boldsymbol{R}_N^{\mathrm{T}}\right]^{\mathrm{T}}, \\ &\boldsymbol{N}_R = \left[(1-\xi)\boldsymbol{I}_3 \quad (1+\xi)\boldsymbol{I}_3\right]^{\mathrm{T}}/2 \end{aligned} \tag{2.2}$$

where N_R is the shape function matrix, q_R is the position vector of the element, $\xi = (2s - s_M - s_N)/(s_N - s_M)$ is a dimensionless material coordinate, and I_3 is the three-size identity matrix. The elemental dynamic equation can be obtained according to the principle of virtual work:

$$\delta W = \int_{s_M}^{s_N} \delta \boldsymbol{R}^{\mathrm{T}} \left(\boldsymbol{F}_{\mathrm{f}} + \boldsymbol{F}_{\mathrm{e}} - \rho A \ddot{\boldsymbol{R}} \right) \mathrm{d}s = 0 \tag{2.3}$$

where $\boldsymbol{F}_{\mathrm{e}}$ is the elastic force, $\boldsymbol{F}_{\mathrm{f}}$ is the external force, and $-\rho A \ddot{\boldsymbol{R}}$ is the inertia force with density ρ and cross-sectional area A. The virtual work of the elastic force can be expressed as

$$\begin{gathered} \boldsymbol{R}^{\mathrm{T}} \boldsymbol{F}_{\mathrm{e}} = \varepsilon E A \left(k\varepsilon + \beta \dot{\varepsilon} \right), \\ \varepsilon = \frac{1}{2} \left(\frac{\partial \boldsymbol{R}^{\mathrm{T}}}{\partial s} \frac{\partial \boldsymbol{R}}{\partial s} - 1 \right), k = \begin{cases} 1, \varepsilon > 0 \\ 0, \varepsilon \le 0 \end{cases} \end{gathered} \tag{2.4}$$

where ε is the axial strain, k indicates that the cable element can only resist tension instead of compression, and β is the Rayleigh damping. Substituting Eqs. (2.2) and (2.4) into Eq. (2.3) yields the governing equations

$$\boldsymbol{M}_{\mathrm{e}} \ddot{q}_{\mathrm{e}} = \boldsymbol{Q}_{\mathrm{m}} + \boldsymbol{Q}_{\mathrm{e}} + \boldsymbol{Q}_{\mathrm{f}} \tag{2.5}$$

in which $\boldsymbol{M}_{\mathrm{e}}$ is the generalized mass matrix of the element, $\boldsymbol{Q}_{\mathrm{m}}$ is the inertial force vector caused by the mass flow, $\boldsymbol{Q}_{\mathrm{e}}$ is the generalized elastic forces and damping forces, and $\boldsymbol{Q}_{\mathrm{f}}$ is the generalized external force vector. These can be calculated through the following equations:

$$\begin{aligned} \boldsymbol{M}_{\mathrm{e}} &= \frac{s_N - s_M}{2} \int_{-1}^{1} \rho A \boldsymbol{N}^{\mathrm{T}} \boldsymbol{N} \mathrm{d}\xi, \\ \boldsymbol{Q}_{\mathrm{m}} &= -\frac{s_N - s_M}{2} \int_{-1}^{1} \rho A \boldsymbol{N}^{\mathrm{T}} \ddot{\boldsymbol{R}} \mathrm{d}\xi, \\ \boldsymbol{Q}_{\mathrm{e}} &= -\frac{s_N - s_M}{2} \int_{-1}^{1} EA \left(\partial \varepsilon / \partial \boldsymbol{q}_{\mathrm{e}} \right)^{\mathrm{T}} \left(k\varepsilon + \beta \dot{\varepsilon} \right) \mathrm{d}\xi, \\ \boldsymbol{Q}_{\mathrm{f}} &= \frac{s_N - s_M}{2} \int_{-1}^{1} \boldsymbol{N}^{\mathrm{T}} \boldsymbol{F}_{\mathrm{f}} \mathrm{d}\xi \end{aligned} \tag{2.6}$$

where the shape function can be expressed as

$$\boldsymbol{N} = \left[\boldsymbol{N}_R \quad \frac{\partial \boldsymbol{N}_R}{\partial s_M} \boldsymbol{q}_R \quad \frac{\partial \boldsymbol{N}_R}{\partial s_N} \boldsymbol{q}_R \right] \tag{2.7}$$

In this formulation, there is no need to keep the length of the ALE element $L_e = |s_N - s_M|$ with the same scale as the pulley size, that is to say, the noncontact cable segment can be meshed roughly; sometimes just one element for each non-contact segment is sufficient.

In the following context, the governing equations for locating the border node of ALE cable elements were formulated. The border location N for the non-sliding case was first derived, then the two border nodes N_A and N_B for the sliding case were derived.

2.3.2 *Border Location of the Non-sliding Case*

A non-sliding case is shown in Fig. 2.2a, in which one of the cable ends F is fixed on the pulley. Assuming that the cable is always tensioned and maintains contact with the pulley, the entire cable FM can be divides into two segments: a contact segment NF and a non-contact segment MN. Their border is the node N of the ALE cable element.

During movement of the cable, the border node N moves in both three-dimensional space and along the cable. However, it obeys several physical rules. First, the border node N must be located on the pulley profile, which means

$$\boldsymbol{C}_1 = \boldsymbol{R}_N - \left[\boldsymbol{R}_P + \boldsymbol{A}(\boldsymbol{\psi})\boldsymbol{r}\right] = \boldsymbol{0} \tag{2.8}$$

where $\boldsymbol{R}_N$ is the global position of the border node N, $\boldsymbol{R}_P$ is the global position of the pulley center, $\boldsymbol{\psi}$ is the Cartesian rotation vector [43] used to describe the attitude of the pulley frame which is fixed on the pulley, $\boldsymbol{A}(\boldsymbol{\psi})$ is the transformation matrix from the pulley frame to the global frame, and $\boldsymbol{r}$ is the local position of node N under the pulley frame. Assuming that the z-axis of the pulley frame $\boldsymbol{z}_P$ is taken along with the axial direction of the pulley and that the cable wraps in the xy-plane only, the local position of node N can be determined by a single parameter φ representing the wrap angle and $\boldsymbol{r}(\varphi) = R\left[\cos\varphi \quad \sin\varphi \quad 0\right]^{\mathrm{T}}$, where R is the pulley radius.

Second, at node N, the non-contact segment of the cable is tangent with the pulley. This yields

$$C_2 = \boldsymbol{t}_N^{\mathrm{T}}\boldsymbol{A}(\boldsymbol{\psi})\boldsymbol{r}(\varphi) = 0 \tag{2.9}$$

where $\boldsymbol{t}_N$ is the tangent of the cable element at its end N.

Finally, the material length of the entire cable, including both the contact and non-contact segments, should be a constant. Here a simple assumption is made that elongation of the contact cable segment is negligible, which is acceptable in mesh antennas as the extension of the contact segment is several orders of magnitude smaller than the extension of the non-contact segment. Then, the above length constraint can be formulated as

$$C_3 = L(s_N, \varphi) - L_0 = sign(d)R\varphi + sign(s_N - s_M)s_N - L_0 = 0 \tag{2.10}$$

with

$$\begin{aligned} & d \triangleq (\boldsymbol{r} \times \boldsymbol{t}_N) \cdot z_P, \\ & sign(x) \triangleq \begin{cases} -1, \text{ if } x < 0 \\ 0, \text{ if } x = 0 \\ 1, \text{ if } x > 0 \end{cases} \end{aligned} \tag{2.11}$$

where L_0 is the initial material length at time t_0, and $d \triangleq (\boldsymbol{r} \times \boldsymbol{t}_N) \cdot z_P$ is introduced to refer to the wrapping direction of the cable.

To sum up, Eqs. (2.8)–(2.10) together offer a set of algebraic constraint equations, five in total, for locating the border node N in the non-sliding case. If $\boldsymbol{q}_P = \left[\boldsymbol{R}_P^{\mathrm{T}}, \boldsymbol{\psi}^{\mathrm{T}}\right]^{\mathrm{T}}$ is introduced to denote the generalized coordinates of the pulley, the constraints are the function of $\boldsymbol{q}_e$, $\boldsymbol{q}_P$, and φ, expressed as

$$\boldsymbol{C}(\boldsymbol{q}_e, \boldsymbol{q}_P, \varphi) = \begin{bmatrix} \boldsymbol{C}_1^{\mathrm{T}} & C_2 & C_3 \end{bmatrix}^{\mathrm{T}} = \boldsymbol{0} \tag{2.12}$$

Here, the wrap angle φ is an intermediate variable rather than a generalized coordinate; it is determined by $\boldsymbol{q}_e$ and $\boldsymbol{q}_P$. In numerical integration, φ can be obtained by solving the tangent constraint $C_2(\boldsymbol{q}_e, \boldsymbol{q}_P, \varphi, t) = 0$ with the Newton–Raphson iteration method [108].

2.3.3 Border Locations and Friction of the Sliding Case

The sliding case is shown in Fig. 2.2b, in which the cable passes through the pulley rather than being fixed on the pulley. At any time, the entire cable $M_A M_B$ contains one contact segment $N_A N_B$ and two noncontact segments $M_A N_A$ and $M_B N_B$, resulting in two border nodes N_A and N_B.

During movement of the cable, both of the two border nodes vary with time, and they obey similar physical rules as in the non-sliding case. That is, nodes N_A and N_B of the cable elements should be located on the pulley, the cables are tangent to the pulley, and the material length of the entire cable is a constant. These conditions are expressed as a set of algebraic constraints:

$$\boldsymbol{C}(\boldsymbol{q}_{e_A}, \boldsymbol{q}_{e_A}, \boldsymbol{q}_P, \boldsymbol{\varphi}) = \begin{bmatrix} \boldsymbol{R}_{N_A} - [\boldsymbol{R}_P + \boldsymbol{A}(\boldsymbol{\psi})\boldsymbol{r}(\varphi_A)] \\ \boldsymbol{R}_{N_B} - [\boldsymbol{R}_P + \boldsymbol{A}(\boldsymbol{\psi})\boldsymbol{r}(\varphi_B)] \\ \boldsymbol{t}_A^{\mathrm{T}} \boldsymbol{A}(\boldsymbol{\psi})\boldsymbol{r}(\varphi_A) \\ \boldsymbol{t}_B^{\mathrm{T}} \boldsymbol{A}(\boldsymbol{\psi})\boldsymbol{r}(\varphi_B) \\ L(s_{N_A}, s_{N_B}, \varphi_A, \varphi_B) - L_0 \end{bmatrix} = \boldsymbol{0} \tag{2.13}$$

where the vector of the wrap angles $\boldsymbol{\varphi} = [\varphi_A, \varphi_B]^{\mathrm{T}}$ is the wrap angle of nodes N_A and N_B respectively, and

$$L = sign(d_A)R\varphi_A + sign(d_B)R\varphi_B + sign\left(s_{N_A} - s_{M_A}\right)s_{N_A} + sign\left(s_{N_B} - s_{M_B}\right)s_{N_B} \quad (2.14)$$

with two wrapping directions $d_A \triangleq (\boldsymbol{r}_A(\varphi_A) \times \boldsymbol{t}_A) \cdot \boldsymbol{z}_P$, $d_B \triangleq (\boldsymbol{r}_B(\varphi_B) \times \boldsymbol{t}_B) \cdot \boldsymbol{z}_P$. The first two items on the right-hand side of Eq. (2.14) represent the contact arc length, while the last two items on the righthand side represent the non-contact length. Again, $\boldsymbol{\varphi} = [\varphi_A, \varphi_B]^{\mathrm{T}}$ can be calculated via the Newton–Raphson iteration method based on the tangent constraints in Eq. (2.13).

Furthermore, for the sliding case, the pulley is usually mounted on a shaft, with or without a bearing. The friction within the bearing or between the pulley and the shaft may result in the decay of cable tension from the active side to the passive side. For convenience, the friction force is represented as

$$f = (1-\eta)T \quad (2.15)$$

where T is the tension of the active side of the cable, and η, where $0 \le \eta \le 1$, is a dimensionless parameter referred to as the transmission efficiency of the cable tension between the passive and the active sides. In practice, η is a function of many physical conditions, such as the working temperature and the cable tension T. However, it is difficult to obtain a theoretical relationship between η and these physical parameters, which are caused by the complex contact details within bearing or between the pulley and the shaft. Therefore, η is measured from experiment when friction forces matter for the studied problem. Here, η is set as a constant for simplicity in the study.

Here an assumption is made that the friction force is exerted equally on the two border nodes N_A and N_B, and hence, the generalized friction forces exerted on the ALE elements e_A and e_B can be calculated according to Eq. (2.6) as

$$\begin{aligned} \boldsymbol{Q}_A^{\mathrm{fri}} &= \frac{1}{2}\begin{bmatrix} \boldsymbol{0}_{3\times1}^{\mathrm{T}} & \boldsymbol{f}_A^{\mathrm{T}} & 0 & -f \end{bmatrix}^{\mathrm{T}}, \\ \boldsymbol{Q}_B^{\mathrm{fri}} &= \frac{1}{2}\begin{bmatrix} \boldsymbol{0}_{3\times1}^{\mathrm{T}} & \boldsymbol{f}_B^{\mathrm{T}} & 0 & -f \end{bmatrix}^{\mathrm{T}} \end{aligned} \quad (2.16)$$

where $\boldsymbol{f}_A$ and $\boldsymbol{f}_B$ are along the tangent direction of the ALE elements e_A and e_B, respectively, and $\|\boldsymbol{f}_A\| = \|\boldsymbol{f}_B\| = f$. Additionally, the friction exerted on the pulley includes the reactive friction force $-(\boldsymbol{f}_A + \boldsymbol{f}_B)/2$ and the friction moment $\boldsymbol{M}_P^{\mathrm{fri}} = \begin{bmatrix} 0 & 0 & -fR \end{bmatrix}^{\mathrm{T}}$ [109]. Therefore, the generalized friction forces exerted on the pulley P can be expressed as

$$\boldsymbol{Q}_P^{\mathrm{fri}} = \begin{bmatrix} -(\boldsymbol{f}_A + \boldsymbol{f}_B)/2 & \left(\boldsymbol{H}(\boldsymbol{\psi})\boldsymbol{M}_P^{\mathrm{fri}}\right)^{\mathrm{T}} \end{bmatrix}^{\mathrm{T}} \quad (2.17)$$

where $\boldsymbol{H}(\boldsymbol{\psi})$ is the so-called tangent operator matrix [43] to transform the local moment vector in the pulley frame to the generalized moment vector, which can be expressed as

$$\boldsymbol{H}(\boldsymbol{\psi}) = \boldsymbol{I}_{3\times3} + \left(\frac{\cos\psi - 1}{\psi^2}\right)\tilde{\boldsymbol{\psi}} + \left(1 - \frac{\sin\psi}{\psi}\right)\frac{\tilde{\boldsymbol{\psi}}\tilde{\boldsymbol{\psi}}}{\psi^2} \tag{2.18}$$

The cross-product matrix [110] $\tilde{\psi}$ can be generated according to the vector $\boldsymbol{\psi} = [\psi_1 \quad \psi_2 \quad \psi_3]^{\mathrm{T}}$ as

$$\tilde{\boldsymbol{\psi}} \triangleq \begin{bmatrix} 0 & -\psi_3 & \psi_2 \\ \psi_3 & 0 & -\psi_1 \\ -\psi_2 & \psi_1 & 0 \end{bmatrix} \tag{2.19}$$

As a result, the generalized vector of friction force and moment can be expressed as

$$\boldsymbol{Q}_{\mathrm{slide}}^{\mathrm{fri}} = \left[\left(\boldsymbol{Q}_A^{\mathrm{fri}}\right)^{\mathrm{T}} \quad \left(\boldsymbol{Q}_B^{\mathrm{fri}}\right)^{\mathrm{T}} \quad \left(\boldsymbol{Q}_P^{\mathrm{fri}}\right)^{\mathrm{T}}\right]^{\mathrm{T}} \tag{2.20}$$

2.3.4 Governing Equations of a Cable-Pulley System

Taking the sliding cable-pulley system as an example, we derive the governing equations according to the Lagrangian equations of the first kind for constrained systems. This system contains a pulley P, two ALE cable elements e_A and e_B, and one virtual contact segment with two wrap angles φ_A and φ_A. The system's generalized coordinate vector is $\boldsymbol{q}_{\mathrm{s}} = \left[\boldsymbol{q}_{e_A}^{\mathrm{T}}, \boldsymbol{q}_{e_B}^{\mathrm{T}}, \boldsymbol{q}_P^{\mathrm{T}}\right]^{\mathrm{T}}$. The vector of the wrap angles $\boldsymbol{\varphi} = [\varphi_A, \varphi_B]^{\mathrm{T}}$ serves as the intermediate variables. They are implicit functions of $\boldsymbol{q}_{\mathrm{s}}$, thus should be eliminated from the governing equations through solving the tangent constraint equations. For the convenience of description, hereafter, $\boldsymbol{C}_{\mathrm{t}}(\boldsymbol{q}_{\mathrm{s}}, \boldsymbol{\varphi}) = \boldsymbol{0}$ represent the tangent constraints, and $\boldsymbol{C}_{\mathrm{s}}(\boldsymbol{q}_{\mathrm{s}}, \boldsymbol{\varphi}) = \boldsymbol{0}$ represent the others.

According to the chain rule of derivation, expressing the variations of $\boldsymbol{\varphi} = \boldsymbol{\varphi}(\boldsymbol{q}_{\mathrm{s}})$ in terms of $\boldsymbol{q}_{\mathrm{s}}$ based on $\boldsymbol{C}_{\mathrm{t}}(\boldsymbol{q}_{\mathrm{s}}, \boldsymbol{\varphi}) = \boldsymbol{0}$ yields

$$\frac{\mathrm{d}\boldsymbol{\varphi}}{\mathrm{d}\boldsymbol{q}_{\mathrm{s}}} = -\left(\frac{\partial \boldsymbol{C}_{\mathrm{t}}}{\partial \boldsymbol{\varphi}}\right)^{-1} \frac{\partial \boldsymbol{C}_{\mathrm{t}}}{\partial \boldsymbol{q}_{\mathrm{s}}} \tag{2.21}$$

Substituting Eq. (2.21) into the Jacobian matrix of the constraints $\boldsymbol{C}_{\mathrm{s}}(\boldsymbol{q}_{\mathrm{s}}, \boldsymbol{\varphi}) = \boldsymbol{0}$ results in

$$\frac{\partial \boldsymbol{C}_{\mathrm{s}}(\boldsymbol{q}_{\mathrm{s}},\boldsymbol{\varphi})}{\partial \boldsymbol{q}_{\mathrm{s}}}=\frac{\partial \boldsymbol{C}_{\mathrm{s}}}{\partial \boldsymbol{q}_{\mathrm{s}}}+\frac{\partial \boldsymbol{C}_{\mathrm{s}}}{\partial \boldsymbol{\varphi}}\frac{\mathrm{d}\boldsymbol{\varphi}}{\mathrm{d}\boldsymbol{q}_{\mathrm{s}}}=\frac{\partial \boldsymbol{C}_{\mathrm{s}}}{\partial \boldsymbol{q}_{\mathrm{s}}}-\frac{\partial \boldsymbol{C}_{\mathrm{s}}}{\partial \boldsymbol{\varphi}}\left(\frac{\partial \boldsymbol{C}_{\mathrm{t}}}{\partial \boldsymbol{\varphi}}\right)^{-1}\frac{\partial \boldsymbol{C}_{\mathrm{t}}}{\partial \boldsymbol{q}_{\mathrm{s}}} \tag{2.22}$$

Therefore, the governing equations of a sliding cable-pulley system can be obtained as a set of differential algebraic equations (DAEs):

$$\begin{cases} \boldsymbol{M}_{\mathrm{s}}\ddot{\boldsymbol{q}}_{\mathrm{s}}+\left(\dfrac{\partial \boldsymbol{C}_{\mathrm{s}}}{\partial \boldsymbol{q}_{\mathrm{s}}}-\dfrac{\partial \boldsymbol{C}_{\mathrm{s}}}{\partial \boldsymbol{\varphi}}\left(\dfrac{\partial \boldsymbol{C}_{\mathrm{t}}}{\partial \boldsymbol{\varphi}}\right)^{-1}\dfrac{\partial \boldsymbol{C}_{\mathrm{t}}}{\partial \boldsymbol{q}_{\mathrm{s}}}\right)^{\mathrm{T}}\boldsymbol{\lambda}_{\mathrm{s}}-\boldsymbol{Q}_{\mathrm{s}}=\boldsymbol{0} \\ \boldsymbol{C}_{\mathrm{s}}(\boldsymbol{q}_{\mathrm{s}},\boldsymbol{\varphi})=\boldsymbol{0} \end{cases} \tag{2.23}$$

where $\boldsymbol{M}_{\mathrm{s}}$ is the assembled mass matrix, $\boldsymbol{Q}_{\mathrm{s}}$ is the vector of generalized forces including the friction force $\boldsymbol{Q}_{\mathrm{slide}}^{\mathrm{fri}}$, and $\boldsymbol{\lambda}_{\mathrm{s}}$ is the vector of the Lagrange multipliers.

2.3.5 Multiscale Modeling Methods of Cable-Cam Systems

The multiscale modeling method of the cable-pulley system has been proposed. It is natural to think of a question: if the circular profile of the pulley is changed to the cam profile, will the multiscale modeling method still be applicable?

The answer is YES. This section is to make an extension from the multiscale modeling of the cable-pulley system to the cable-cam system. In fact, the cable-cam system is common in life. For example, the composite bow utilizes the cable-cam system to achieve an output of the nonlinear force, and a passive actuator with non-linear force-displacement curves can also meet the specific demand of unfolding mesh antennas [111]. Besides, the cable-cam system can also be applied to the design of gravity compensation mechanisms. Therefore, it is of great significance for the modeling and simulation of the cable-cam system.

Essentially, the difference between a pulley and a cam is the profile. The circular profile of the pulley can be described by the center and radius, while the description of the cam profile is more complicated. Generally, the cam profile can be interpolated according to a finite number of discrete nodes, and the interpolation curve is required to keep the original geometry properties, such as the shape-preserving property. However, general interpolation methods, such as the Hermite interpolation, cannot ensure a convex curve based on the interpolation of a series of convex nodes. Fang [112] constructs a class of parametric spline curve which interpolates all the discrete nodes and has C2 continuity. Considering the contact between the cable and the cam, C1 continuity is enough for the interpolation. Based on the above parametric spline curve, this section develops the cubic spline curve with convexity-preserving and C1-continuity properties.

For simplicity, the cam profile can be described by parametric equations, and the definition of parametric coordinate u is shown in Fig. 2.3b. The parameter

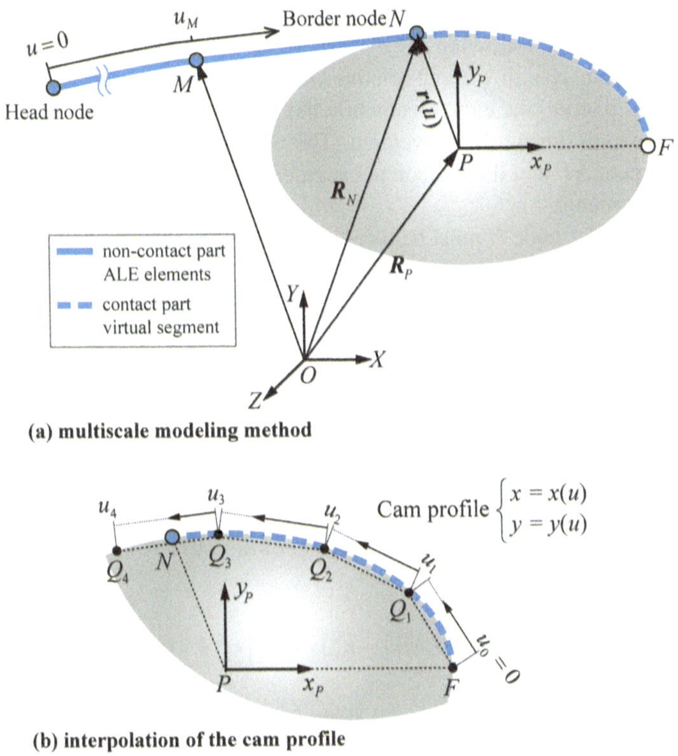

Fig. 2.3 Multiscale modeling of the cable-cam system: (**a**) multiscale modeling method; (**b**) interpolation of the cam profile

coordinate at each interpolation node u_i is the sum length of the line segment from the starting node F to the current node as

$$u_i = \begin{cases} 0, i = 0 \\ \sum_{i=1}^{i} \left| \overline{Q_{i-1}Q_i} \right|, i = 1, 2, \cdots, n+1 \end{cases} \tag{2.24}$$

The parametric coordinate u at any point of the cam profile can be obtained by a linear interpolation. The parametric equations of the cam profile can be expressed as

$$\begin{cases} x = x(u) \\ y = y(u) \end{cases} \tag{2.25}$$

Similarly, the non-contact part of the cable is meshed using variable-length cable elements based on the ALE formulation, illustrated by the solid lines; the contact part is considered as a virtual segment whose length is time variant, as illustrated by the dashed line. The border node between the non-contact segment and contact segment is located in the following.

The cable-cam system without relative sliding is shown in Fig. 2.3a. One end of the cable F is fixed to the cam, meanwhile the cable is tensioned and keeps in contact with the cam during the movement. The whole cable can be divided into the contact segment NF and the noncontact segment MN, and N is the dynamic border of the two segments.

First, the border node N must be located on the cam profile, which means

$$\boldsymbol{C}_1 = \boldsymbol{R}_N - \left[\boldsymbol{R}_P + \boldsymbol{A}(\boldsymbol{\psi})\boldsymbol{r}(u)\right] = \boldsymbol{0} \tag{2.26}$$

where $\boldsymbol{R}_N$ is the global position of the border node N, $\boldsymbol{R}_P$ is the global position of the cam center, $\boldsymbol{\psi}$ is the Cartesian rotation vector used to describe the attitude of the cam frame which is fixed on the cam, $\boldsymbol{A}(\boldsymbol{\psi})$ is the transformation matrix from the cam frame to the global frame, and $\boldsymbol{r}(u) = \left[x(u) \quad y(u) \quad 0\right]^{\mathrm{T}}$ is the local position of node N under the cam frame.

Second, at node N, the non-contact segment of the cable is tangent with the cam. This yields

$$C_2 = \boldsymbol{A}^{\mathrm{T}}(\boldsymbol{\varphi}_P)(\boldsymbol{r}_N - \boldsymbol{r}_M) \times \boldsymbol{r}_N' = [* \quad * \quad 0]^{\mathrm{T}} \tag{2.27}$$

in which $\boldsymbol{r}'(u)$ is the tangent of the cam at the border, $\boldsymbol{A}^{\mathrm{T}}(\boldsymbol{\varphi}_P)(\boldsymbol{r}_N - \boldsymbol{r}_M)$ is the tangent of the cable element at the border. The parallelism of the $\boldsymbol{r}'(u)$ and $\boldsymbol{A}^{\mathrm{T}}(\boldsymbol{\varphi}_P)(\boldsymbol{r}_N - \boldsymbol{r}_M)$ can be mathematically expressed as the third component of the cross product equals zero. It is worth mentioning that, different from the cable-pulley system, the radial vector $\boldsymbol{r}(u)$ is no longer perpendicular to the cable.

Finally, when the elongation of the contact cable segment is negligible, the above length constraint can be formulated as

$$\begin{aligned} C_3 = L(s_N, u) - L_0 = sign(d)\int_0^s \sqrt{x'^2(u) + y'^2(u)}\mathrm{d}u \\ + sign(s_N - s_M)s_N - L_0 = 0 \end{aligned} \tag{2.28}$$

where $d \triangleq (\boldsymbol{r}_N \times \boldsymbol{r}_N') \cdot \boldsymbol{z}_P$.

To sum up, Eqs. (2.26)–(2.28) together offer a set of algebraic constraint equations, five in total, in the non-sliding case. If $\boldsymbol{q}_P = \left[\boldsymbol{R}_P^{\mathrm{T}}, \boldsymbol{\psi}^{\mathrm{T}}\right]^{\mathrm{T}}$ is introduced to denote the generalized coordinates of the cam, the constraints are the function of $\boldsymbol{q}_e$, $\boldsymbol{q}_P$, and u, expressed as

$$\boldsymbol{C}(\boldsymbol{q}_e, \boldsymbol{q}_P, s) = \left[\boldsymbol{C}_1^{\mathrm{T}} \quad C_2 \quad C_3\right]^{\mathrm{T}} = \boldsymbol{0} \tag{2.29}$$

Here, the parameter coordinate u is an intermediate variable rather than a generalized coordinate; it is determined by $\boldsymbol{q}_e$ and $\boldsymbol{q}_P$. In numerical integration, u can be obtained by solving the tangent constraint $C_2(\boldsymbol{q}_e, \boldsymbol{q}_P, u, t) = 0$ with the Newton–Raphson iteration method. Similar to the cable-pulley system, the governing equation of the cable-cam system of the non-sliding is the same as that given in Eq. (2.23). For the sliding case of the cable-cam system, the derivation of the governing equation is similar to that of the cable-pulley system, which will not be repeated.

2.3.6 *Multiscale Modeling Validations*

In this section, three examples are used to verify the correctness of the multi-scale modeling method for the cable-pulley/cam system and illustrate the advantages of the multi-scale modeling method.

2.3.6.1 Yo-Yo Ball

A Yo-Yo ball is a small toy which is popular around the world. It consists of a pair of thin disks and a thin-walled central cylinder, as shown in Fig. 2.4a. One end of the cable is wrapped around the central cylinder while the other end of the cable is fixed on the fingers. When the Yo-Yo ball is released, it spins against the winding direction and falls vertically until the string unfolded, then it turns and climbs up again. The movement process of the Yo-Yo ball is as follows: falling-turning-climbing. Given that the mass of the Yo-Yo is m, the moment of inertia relative to the cylindrical axis is I, the radius of the thin-walled central cylinder is R, the gravitational acceleration is g, and the density, cross-sectional area and elastic modulus of the cable are respectively expressed by ρ, A, E. These parameters are listed in Table 2.1.

In this example, the multiscale modeling method and the conventional contact modeling method were respectively used to model the Yo-Yo ball, as shown in Fig. 2.4b. For the multiscale model, the non-contact segment of the cable was modeled by an ALE cable element, and the tangential point between the string and the central cylinder acted as a dynamic boundary. For the contact model, the whole cable was discretized into several fixed-length cable elements, the contact between the node of these elements and the contour of the central cylinder was defined, and the collision force was applied. In addition, if the elongation of the cable is ignored, an analytical solution can be obtained to describe the Yo-Yo ball. The detailed solution can be seen in reference [113].

The simulation results and analytical solution of the two models were shown in Fig. 2.4c, which includes the falling distance, wrapping angle and tension of the cable during the whole falling-turning-climbing process. For the contact model, it can be seen that the tension oscillation caused by the contact modeling method resulted in the small simulation step and low calculation efficiency. As a contrast, the multiscale modeling method ensures high calculation efficiency since the contact is avoided. However, considering the numerical damping of backward

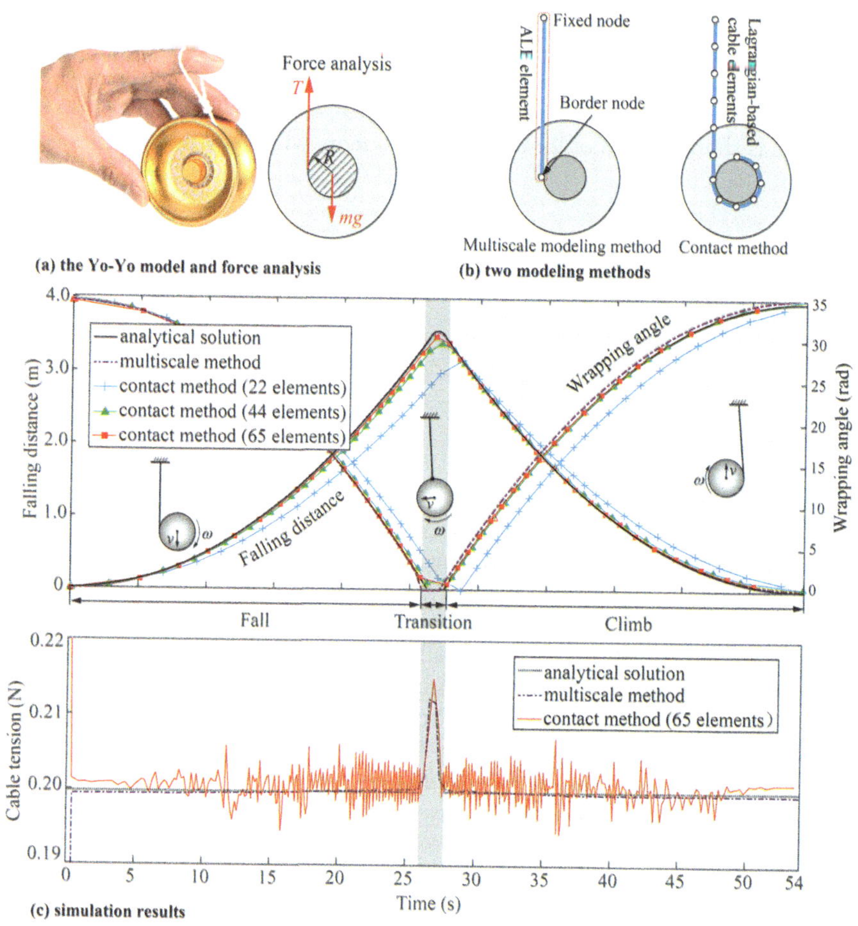

Fig. 2.4 Modeling and simulation results of the Yo-Yo ball: (**a**) the Yo-Yo model and force analysis; (**b**) two modeling methods; (**c**) simulation results

Table 2.1 Parameters of the Yo-Yo ball model

Density ρ (kg/m^{-3})	Elastic modulus E (Pa)	Cross-sectional area of the cable A (m^2)	Mass of the Yo-Yo ball m (kg)	Moment of inertia of the Yo-Yo ball I (kg/m^2)	Radius of the thin-walled central cylinder R (m)
14.4	4×10^7	$\pi \times 10^{-6}$	0.02	0.2	0.1

difference method (BDF), the multiscale method cannot guarantee the energy conservation of the system.

Table 2.2 compares the number of generalized coordinates, relative error and CPU time between the multiscale and contact model. It can be seen that the contact method needs to divide the cable into dense units to obtain better accuracy. For

Table 2.2 Comparison between the multiscale model and the contact model

Modeling method		Number of generalized coordinates	Relative errors (%)	CPU time
Multiscale method	1 ALE element	20	0.53	9
Contact method	22 element	84	18.57	75
	44 element	150	4.39	113
	65 element	216	1.58	139

example, when the cable is only divided into 22 elements, the error relative to the analytic solution is 18.57%; when the string is divided into 65 units, the relative error reduces to 1.58%. For the multiscale model, only one ALE element is needed with the relative error less than 0.53%. It can be seen that the reduction of the number of generalized coordinates and the avoidance of contact detection greatly improve the computational efficiency of the multi-scale method.

2.3.6.2 Constant Force Design by Involute Cam

As shown in Fig. 2.5a, a constant force mechanism can be realized by using the cam-cable system. The involute cam in the vertical plane can rotate around the node O, the cable CD and tension spring with the elastic coefficient of k are connected in series. One end is wrapped around the cylinder of the cam with radius of R, the other end is fixed on the ground. One end of the cable AB is wrapped around the involute cam, and the other end is connected to a weight of mass m. If the radius of the base circle of the involute is r, the equation of the involute cam in the global system at the initial moment can be expressed as

$$\begin{cases} x = r(\cos\varphi + \varphi\sin\varphi) \\ y = r(\sin\varphi - \varphi\cos\varphi) \end{cases} \tag{2.30}$$

The horizontal distance between the tangent node B and the center of the circle O is set as x, and the factors of uneven CAM inertia are ignored. According to the geometric relation and equilibrium condition [114], the relation to be satisfied is

$$r = \frac{R^2 k}{mg} \tag{2.31}$$

The spring tension at the initial time is denoted as T_0, and the horizontal distance between the tangent point B and the center of the circle O is denoted as x_0. The gravity compensation requires that

$$T_0 R = mgx_0 \tag{2.32}$$

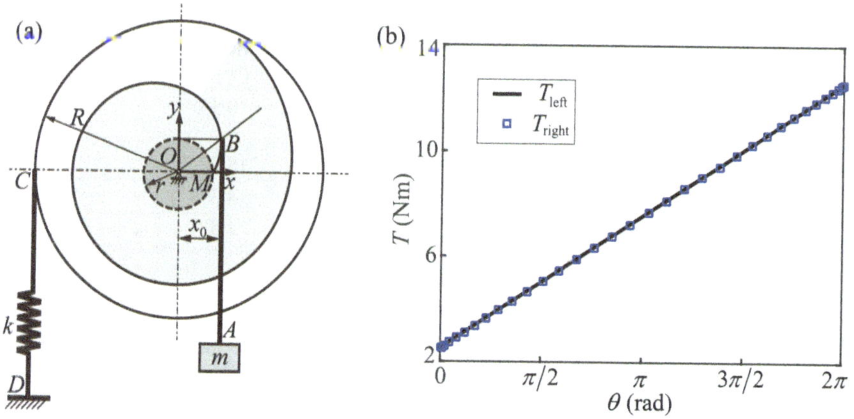

Fig. 2.5 Design of the constant force mechanism: (**a**) involute cam-linear spring; (**b**) simulation result

The multiscale model established in this example was as follows: the cam was modeled as a rigid body, the cable *AB* and the cable *CD* were discretized by ALE cable element, and the tangent node to the cam acted as the dynamic boundary. The cable *CD* was connected in series with a linear spring with stiffness *K*. The rotation angle of the cam relative to the global coordinate frame is θ, and the torque of the cable *CD* and the cable *AB* by the cam are T_{left} and T_{right} respectively. The simulation results were shown in Fig. 2.5b. It can be seen that no matter how much angle the cam rotated, the torques on the left and right sides always remained the same. Under the above conditions, the cam can achieve random equilibrium, and the vertical upward tension provided by the right cable *AB* is always *mg*. The example also confirms the correctness of the multiscale modeling method of the cable-cam system.

2.3.6.3 Two-Bay Model of a Mesh Antenna

The third verification examples is a two-bay model of the 12.25 m Astromesh-II, as shown in Fig. 2.6a, which contains multiple cable-pulley systems. It is also the basic structure of the next section. The relevant parameters of the model are given in Fig. 2.6 and Table 2.3. The longerons, battens, and diagonals were meshed with geometrically exact beam elements [115]. The connections, including T-shaped connections, fixed connections, and translational connections, were regarded as rigid bodies. Additionally, several types of constraints were adopted, including revolute joints, fixed joints, and flexible translational joints along battens. The latch mechanism for translational joints can be achieved with nonlinear forces. A smooth motion of displacement was applied to each translational joint to fold the two-bay model as

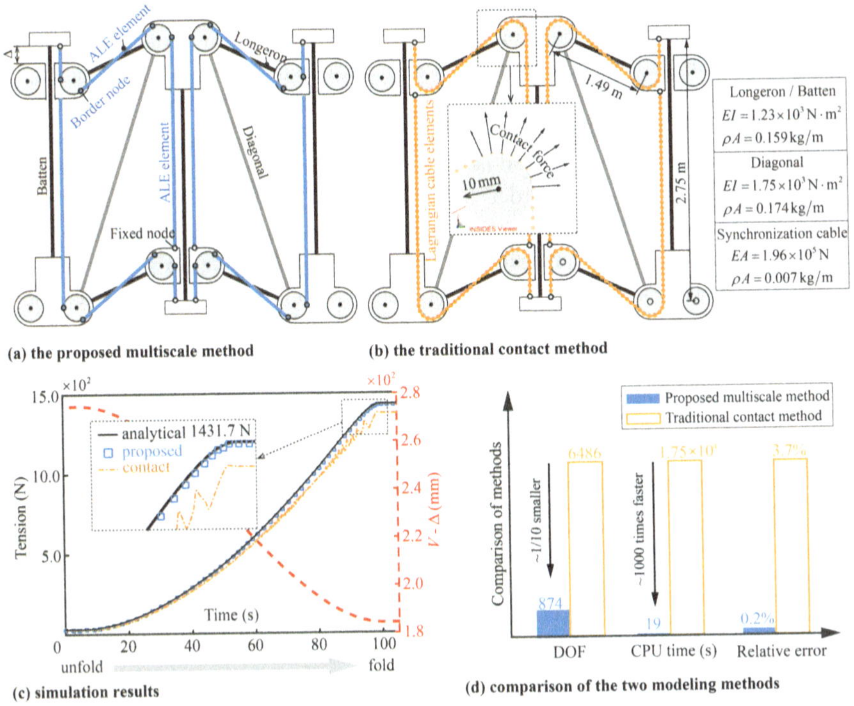

Fig. 2.6 Comparison of two modeling methods for a two-bay truss: (**a**) the proposed multiscale method; (**b**) the traditional contact method; (**c**) simulation results; (**d**) comparison of the two modeling methods. Adapted with permission from [107]

Table 2.3 Parameters of the two-bay model

	Bending stiffness EI (N·m²)	Linear density ρA (kg/m)	Tensile stiffness (N)
Longeron/batten	1.23×10	0.159	–
Diagonal	1.75×10^3	0.174	–
Synchronization cable	–	0.007	1.96×10^5

$$\Delta = \mathrm{STEP}(t, t_1, \Delta_1, t_2, \Delta_2) \triangleq \begin{cases} \Delta_1 & t \le t_1 \\ \Delta_1 + (\Delta_2 - \Delta_1)3 - 2\overline{t}^2 & t_1 < t < t_2 \\ \Delta_2 & t \ge t_2 \end{cases} \tag{2.33}$$

where $\overline{t} = (t - t_1)/(t_2 - t_1)$. The parameters include $t_1 = 5$ s, $t_2 = 100$ s, $\Delta_1 = 0$ mm, $\Delta_2 = 809$ mm.

Two modeling methods were compared: the traditional contact method and the proposed multiscale method. For the contact method as shown in Fig. 2.6b, the synchronization cables were meshed with Lagrangian-based cable elements, in which the elemental length was scaled to the pulley size for valid contact. However, for the multiscale method, only one ALE element for each non-contact segment was needed to obtain a converged and accurate result, since the tension in the cable is uniformly distributed. The border locations were determined by the governing equations for sliding cable pulley systems as shown in Eq. (2.23). Both the contact and the proposed models were built and simulated based on our lab's in-house multibody code [116]. Additionally, the tension of the synchronization cable was calculated via $T = EA\Delta l/l$, where E, A and l denote the Young's modulus, cross-sectional area, and the original length of the synchronization cable respectively, and Δl is defined as the extension of the synchronization cable.

An overall comparison of the above two modeling methods is given in Fig. 2.6c and d. In terms of accuracy, the results for the tensions of the synchronization cables from the deployed state to the stowed state show that the proposed method coincides quite well with the analytical solution, while there is notable oscillation in the contact model with a maximal difference of 54 N relative to the analytical solution. In terms of efficiency, the calculation speed of the proposed method is nearly 1000 times faster than the time needed for the traditional contact method. This increase in calculation speed is caused not only by the fact that the number of generalized coordinates is cut down by 85% but also by the fact that the proposed method bypasses the evaluation of contact forces, which requires small integration steps and reduces the calculation speed.

2.4 Multibody Dynamic Model of a Mesh Antenna

As shown in Fig. 2.7, a mesh antenna consists of five parts: a ring truss, a front net, a rear net, a metal mesh and an array of tie springs. Some parts are fixed to each other, and some allow relative movement, that is, there exist different types of constraints. In addition, the array of tie springs, the friction of the cables across the pulleys, and the locking of the sliding joints when the antenna is fully unfolded can be achieved by applying loads. In the process of unfolding, most components will undergo a wide range of spatial translation and rotation, and the ring truss and the cable networks may suffer flexible deformation, which is in line with the typical research category of flexible multibody dynamics.

In this section, the multibody dynamic model of the deployable mesh antenna Astromesh-II is established, which can be classified into three types including rigid/flexible body elements, constraints and loads. (1) There are 4 types of elements, as shown in Table 2.4: fixed joints, T-joints, sliders and pulleys were modeled by rigid body. Longerons, battens and diagonals were modeled by geometrically accurate beam element. The front net and the rear net were modeled by the Lagrangian cable

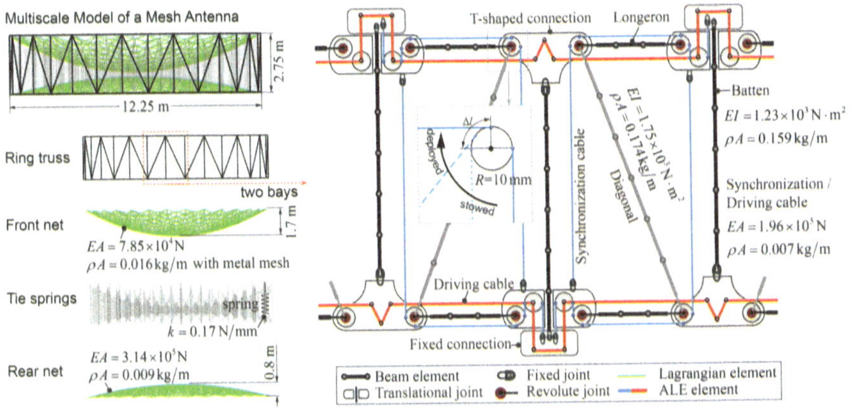

Fig. 2.7 Multibody dynamic model of the deployable mesh antenna. Adapted with permission from [107]

Table 2.4 Types of the elements

No.	Type	Component
1	Rigid	Fixed connection, T-shaped connection, slider, pulley
2	Geometrically accurate beam	Longeron, batten, diagonal
3	Lagrangian cable element	Front net, rear net
4	ALE cable element	Driving cable, synchronization cable

Table 2.5 Types of the constraints

No.	Type	Location
1	Spherical	Between front/rear nets and battens
2	Fixed	Between fixed/T-shaped connections and battens
3	Revolute	Between fixed connections and longerons/battens, and between T-shaped connections and longerons/battens
4	Flexible translational	Between sliders and battens

element. The synchronization cables and the driving cables were modeled by the ALE cable element. (2) There are 4 types of constraints, as shown in Table 2.5: the spherical joints between the front/rear nets and the battens; the fixed joints between the fixed/T-shaped connections and the battens; the revolute joints between the fixed connections and the longerons/battens, and between the T-shaped connections and the longerons/battens; the flexible translational joints between the sliders and the battens. (3) There are 4 types of loads as shown in Table 2.6: the locking force of the sliders when each bay is fully unfolded, the tension force provided by the tie springs between the front net and the rear net, the friction force of the driving cables through the pulleys, and the friction force of the sliders along the battens.

Table 2.6 Types of the loads

No.	Type	Description
1	Locking force	The sliders are locked when each bay is fully unfolded
2	Tension force	Tie springs between the front net and the rear net
3	Friction force	Driving cables through pulleys
4	Friction force	Sliders along battens

In view of the existing modeling theories and methods of rigid body, geometrically accurate beam and the constraints including spherical and fixed joints, the governing equations of the cable-pulley system have been derived in the last section for modeling the synchronization cable-pulley system. The modeling methods will be described as follows, such as the tension decay of the driving cables through the pulleys, the flexible translational joints, and the stick-slip friction.

2.4.1 *Tension Decay of Driving Cables Through Pulleys*

For Astromesh-II, the ring truss is deployed by reeling in the two driving cables, which are controlled by four motors. Two motors are fixed at the upward batten of the first bay while the other two are fixed at the downward batten. The four motors simultaneously begin to shorten the two ends of the up-driving and down-driving cables with reel-in velocity

$$v_{\text{motor}} = 20 - 0.01T \tag{2.34}$$

where T represents the cable tension exerted on the motor, also called the driving force. The velocity decreases as the driving force increases, a relationship determined by the electrical-mechanical properties of the motors. The unit of vm is millimeters per second and the unit of T is Newtons.

The length of the driving cable is gradually shortened during the unfolding process. For the driving cables, the size of the pulleys through which they pass is not important for deployment dynamics, so we considered those pulleys as dimensionless points. The cable networks were meshed with Lagrangian cable elements [117]. In Sect. 2.3.1, the governing equation of ALE cable element with time-varying length is derived, in which the node of the element is free, and no constraint conditions are imposed on the node position and material coordinates. For the driving cable-pulley system, the size effect of pulley is ignored and regarded as a point. If the governing equation of driving cable is established with ALE cable element, it is only necessary to establish the constraint equation of ALE node and apply the generalized force corresponding to the friction force.

As the pulley is simplified as a point, the ALE element node M which keeps in contact with pulley P needs to be satisfied

$$\boldsymbol{R}_M - \boldsymbol{R}_P = \boldsymbol{0} \tag{2.35}$$

where $\boldsymbol{R}_M$ and $\boldsymbol{R}_P$ are the positions of ALE element node M and the pulley center P respectively. Similarly, considering the tension decay caused by the friction between the driving cable and the pulley, the friction force $f = (1 - \eta)T$ is applied to the ALE node which keeps in contact with the pulley. Peng Yun et al. [31] have proved that the generalized force corresponding to the friction force only needs to be applied on the material coordinates of ALE node, that is, the generalized force corresponding to the friction force of the driving cable through the pulley can be expressed as

$$\boldsymbol{Q}_{\mathrm{f}} = \begin{bmatrix} \boldsymbol{0}_{1\times 3} & -f \end{bmatrix}^{\mathrm{T}} \tag{2.36}$$

For the ALE node N which is connected to the motor, its constraints need to meet two requirements: Firstly, the position of node N needs to be consistent with the position of the motor; Secondly, taking into account the reel-in cable by motor, the node N is accompanied by the flow of substances. To sum up, the constraint equation of ALE node N which is connected to the motor can be expressed as

$$\begin{aligned} \boldsymbol{R}_N - \boldsymbol{R}_{\mathrm{motor}} &= \boldsymbol{0}, \\ \dot{s}_N - v_{\mathrm{motor}} &= 0 \end{aligned} \tag{2.37}$$

where $\boldsymbol{R}_N$ and $\boldsymbol{R}_{\mathrm{motor}}$ are the positions of ALE node N and the motor respectively. $\dot{s}_N$ is the change rate with time of material coordinates at node N, which reflects the material velocity at node N.

2.4.2 Modeling of Flexible Sliding Joints

For modeling the flexible sliding joints, the constraint equations are formulated to avoid the traditional contact detection and improve the calculation efficiency. Here a brief description of the constraint equations is given.

As shown in Fig. 2.8, the slider is moving along the flexible beam. The position and attitude of the node of the sliding body in the global frame $O_g x_g y_g z_g$ can be expressed as $(\boldsymbol{r}_s, \boldsymbol{\varphi}_s)$, the sliding coordinate frame Marker s is fixed on the sliding body and slides along the sliding curve; The sliding curve is determined by two local coordinate frames, Marker 1 and Marker 2, which are fixed on the two nodes N_1 and N_2 respectively. $(\boldsymbol{r}_{b1}, \boldsymbol{\varphi}_{b1})$ and $(\boldsymbol{r}_{b2}, \boldsymbol{\varphi}_{b2})$ describe the positions and attitudes of the nodes N_1 and N_2 respectively.

The global position of Marker i ($i = 1, 2$) in the coordinate frame of sliding curve can be expressed as

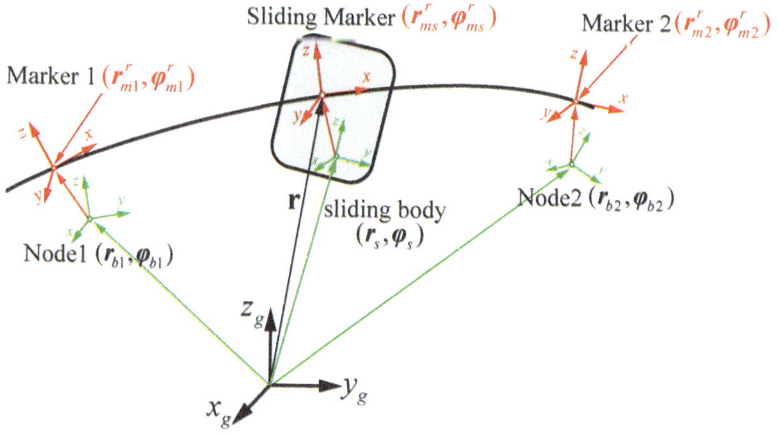

Fig. 2.8 Modeling of the flexible sliding joint

$$\boldsymbol{r}_{mi} = \boldsymbol{r}_{bi} + \boldsymbol{A}(\boldsymbol{\varphi}_{bi})\boldsymbol{r}_{mi}^{r}, i = 1,2 \tag{2.38}$$

The position vector $\boldsymbol{r}$ of the sliding node on the sliding curve can be obtained by Hermite interpolation as

$$\boldsymbol{r} = \sum_{i=1}^{4} N_i \boldsymbol{r}_{mi} = N_1 \boldsymbol{r}_{m1} + N_2 \boldsymbol{X}_{m1} + N_3 \boldsymbol{r}_{m2} + N_4 \boldsymbol{X}_{m2} \tag{2.39}$$

in which the shape function

$$\begin{aligned} N_1 &= 2s^3 - 3s^2 + 1 \\ N_2 &= L_0\left(s^3 - 2s^2 + s\right) \\ N_3 &= -2s^3 + 3s^2 \\ N_4 &= L_0\left(s^3 - s^2\right) \end{aligned} \tag{2.40}$$

L_0 is the arc length of the sliding segment; s is the locally normalized internal parameter, which represents the position of the sliding node in the sliding segment.

The rotation vector $\boldsymbol{\varphi}_{\text{relb}}$ represents the attitude transformation from left Marker 1 to the right Marker 2

$$\boldsymbol{A}(\boldsymbol{\varphi}_{b1})\boldsymbol{A}(\boldsymbol{\varphi}_{m1}^{r})\boldsymbol{A}(\boldsymbol{\varphi}_{\text{relb}}) = \boldsymbol{A}(\boldsymbol{\varphi}_{b2})\boldsymbol{A}(\boldsymbol{\varphi}_{m2}^{r}) \tag{2.41}$$

The attitude of the sliding point on the sliding curve relative to the left Marker 1 can be obtained by linear interpolation as

$$\boldsymbol{\varphi}_{\text{rels}}^{\text{interp}} = s\boldsymbol{\varphi}_{\text{relb}} \tag{2.42}$$

The attitude matrix $\Lambda_s^{\text{interp}}$ of the sliding node on the sliding curve can be obtained by interpolation as

$$\boldsymbol{\Lambda}_s^{\text{interp}} = \left[\boldsymbol{i}_s^{\text{interp}} \quad \boldsymbol{j}_s^{\text{interp}} \quad \boldsymbol{k}_s^{\text{interp}} \right] = \boldsymbol{\Lambda}(\boldsymbol{\varphi}_{b1})\boldsymbol{\Lambda}(\boldsymbol{\varphi}_{m1}^{r})\boldsymbol{\Lambda}(\boldsymbol{\varphi}_{\text{rels}}^{\text{interp}}) \tag{2.43}$$

The attitude matrix of sliding mark points on the slider can be expressed as

$$\boldsymbol{\Lambda}_s = \boldsymbol{\Lambda}_s^{\text{interp}} = [\boldsymbol{i}_s \quad \boldsymbol{j}_s \quad \boldsymbol{k}_s] = \boldsymbol{\Lambda}(\boldsymbol{\varphi}_s)\boldsymbol{\Lambda}(\boldsymbol{\varphi}_{ms}^{r}) \tag{2.44}$$

Figure 2.9 shows three types of flexible sliding joints, including spherical joints, cylindrical joints and translational joints. Spherical joint requires that the position of the sliding node is located on the sliding curve, hence only translational degree of freedom is constrained. Cylindrical joint requires that the position of the sliding node is located on the sliding curve, and the sliding direction of the slider coincides with the tangential direction of the sliding curve. For the cylindrical joint, both the translational degree of freedom and two rotational degrees of freedom are limited. Translational joint requires that the sliding node of the slider is located on the sliding curve and the attitude of the slider is consistent with the sliding curve, hence both the translational and rotational degrees of freedom are limited.

If the sliding joint belongs to the spherical type, only the position of the sliding node is constrained. Therefore, the constraint equation can be expressed as

$$\boldsymbol{r} - \left(\boldsymbol{r}_s + \boldsymbol{\Lambda}(\boldsymbol{\varphi}_s)\boldsymbol{r}_{ms}^{r}\right) = \boldsymbol{0} \tag{2.45}$$

If the sliding joint belongs to the cylindrical type, the x-axis of the sliding coordinate frame on the slider should be perpendicular to the y-axis and z-axis of the sliding curve obtained by interpolation. The attitude constraint equation should be added to Eq. (2.45) as

$$\begin{cases} \dfrac{\partial \boldsymbol{r}}{\partial s} \cdot \boldsymbol{j}_s = 0 \\ \dfrac{\partial \boldsymbol{r}}{\partial s} \cdot \boldsymbol{k}_s = 0 \end{cases} \tag{2.46}$$

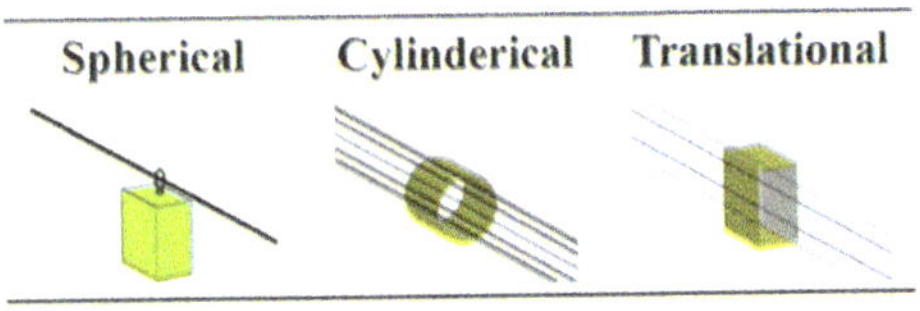

Fig. 2.9 Three types of the flexible sliding joints

If the sliding joint belongs to the translational type, the y-axis of the sliding coordinate frame on the slider should be perpendicular to the z-axis of the sliding curve obtained by interpolation. The attitude constraint equation is added to Eqs. (2.45) and (2.46) as

$$\boldsymbol{j}_s \cdot \boldsymbol{k}_s^{\text{interp}} = 0 \tag{2.47}$$

Taking the flexible translational joint as an example, the constraint equation can be written as follows when $\boldsymbol{A}(\boldsymbol{\varphi}_{b1})\boldsymbol{A}(\boldsymbol{\varphi}_{m1}^{r}) = [\boldsymbol{i} \quad \boldsymbol{j} \quad \boldsymbol{k}]$

$$\boldsymbol{C} = \begin{bmatrix} \boldsymbol{i} \cdot \left[\boldsymbol{r} - \left(\boldsymbol{r}_s + \boldsymbol{A}(\boldsymbol{\varphi}_s)\boldsymbol{r}_{ms}^{r}\right)\right] \\ \boldsymbol{j} \cdot \left[\boldsymbol{r} - \left(\boldsymbol{r}_s + \boldsymbol{A}(\boldsymbol{\varphi}_s)\boldsymbol{r}_{ms}^{r}\right)\right] \\ \boldsymbol{k} \cdot \left[\boldsymbol{r} - \left(\boldsymbol{r}_s + \boldsymbol{A}(\boldsymbol{\varphi}_s)\boldsymbol{r}_{ms}^{r}\right)\right] \\ \boldsymbol{i}_s \cdot \boldsymbol{j}_s^{\text{interp}} \\ \boldsymbol{i}_s \cdot \boldsymbol{k}_s^{\text{interp}} \\ \boldsymbol{j}_s \cdot \boldsymbol{k}_s^{\text{interp}} \end{bmatrix} = \begin{bmatrix} \boldsymbol{C}_L \\ \boldsymbol{C}_M \end{bmatrix} = \boldsymbol{0} \tag{2.48}$$

where $\boldsymbol{C}_M$ is a column vector with 5 elements. Next, the following can be obtained as

$$\delta \boldsymbol{C} = \begin{bmatrix} \delta \boldsymbol{C}_L \\ \delta \boldsymbol{C}_M \end{bmatrix} = \begin{bmatrix} \boldsymbol{C}_{L,q}\delta \boldsymbol{q} + \boldsymbol{C}_{L,s}\delta s \\ \boldsymbol{C}_{M,q}\delta \boldsymbol{q} + \boldsymbol{C}_{M,s}\delta s \end{bmatrix} = \boldsymbol{0} \tag{2.49}$$

It can be obtained from the first line of Eq. (2.49) as

$$\delta s = -\boldsymbol{C}_{L,s}^{-1}\boldsymbol{C}_{L,q}\delta \boldsymbol{q} = \boldsymbol{G}\delta \boldsymbol{q} \tag{2.50}$$

in which

$$\boldsymbol{G} = -\boldsymbol{C}_{L,s}^{-1}\boldsymbol{C}_{L,q} \tag{2.51}$$

The generalized coordinates $\boldsymbol{q} = \left[\boldsymbol{r}_{b1}^{\mathrm{T}} \quad \boldsymbol{\varphi}_{b1}^{\mathrm{T}} \quad \boldsymbol{r}_{b2}^{\mathrm{T}} \quad \boldsymbol{\varphi}_{b2}^{\mathrm{T}} \quad \boldsymbol{r}_{s}^{\mathrm{T}} \quad \boldsymbol{\varphi}_{s}^{\mathrm{T}}\right]^{\mathrm{T}}$ are selected, and the dynamic equation without constraints can be expressed as $\boldsymbol{M}\ddot{\boldsymbol{q}} - \boldsymbol{Q} = \boldsymbol{0}$. According to the first type of Lagrange equation, the differential equation considering constraints is

$$\delta \boldsymbol{q}^{\mathrm{T}}\left(\boldsymbol{M}\ddot{\boldsymbol{q}} - \boldsymbol{Q}\right) + \delta \boldsymbol{C}^{\mathrm{T}}\boldsymbol{\lambda} = \boldsymbol{0} \tag{2.52}$$

According to $\delta \boldsymbol{C} = \boldsymbol{C}_q\delta \boldsymbol{q} + \boldsymbol{C}_s\delta s$, the following is obtained as

$$\delta \boldsymbol{q}^{\mathrm{T}}\left(\boldsymbol{M}\ddot{\boldsymbol{q}}+\boldsymbol{C}_{q}^{\mathrm{T}}\boldsymbol{\lambda}-\boldsymbol{Q}\right)+\delta s\boldsymbol{C}_{s}^{\mathrm{T}}\boldsymbol{\lambda}=\mathbf{0} \tag{2.53}$$

i.e.,

$$\delta \boldsymbol{q}^{\mathrm{T}}\left(\boldsymbol{M}\ddot{\boldsymbol{q}}+\begin{bmatrix}\boldsymbol{C}_{L,q}^{\mathrm{T}} & \boldsymbol{C}_{M,q}^{\mathrm{T}}\end{bmatrix}\begin{bmatrix}\boldsymbol{\lambda}_L \\ \boldsymbol{\lambda}_M\end{bmatrix}-\boldsymbol{Q}\right)+\delta s\begin{bmatrix}\boldsymbol{C}_{L,s}^{\mathrm{T}} & \boldsymbol{C}_{M,s}^{\mathrm{T}}\end{bmatrix}\begin{bmatrix}\boldsymbol{\lambda}_L \\ \boldsymbol{\lambda}_M\end{bmatrix}=\mathbf{0} \tag{2.54}$$

According to $\delta s = \boldsymbol{G}\delta\boldsymbol{q}$, the following is obtained as

$$\boldsymbol{M}\ddot{\boldsymbol{q}}+\left(\boldsymbol{C}_{L,q}^{\mathrm{T}}+\boldsymbol{G}^{\mathrm{T}}\boldsymbol{C}_{L,s}^{\mathrm{T}}\right)\boldsymbol{\lambda}_L+\left(\boldsymbol{C}_{M,q}^{\mathrm{T}}+\boldsymbol{G}^{\mathrm{T}}\boldsymbol{C}_{M,s}^{\mathrm{T}}\right)\boldsymbol{\lambda}_M-\boldsymbol{Q}=\mathbf{0} \tag{2.55}$$

Considering that $\boldsymbol{G}=-\boldsymbol{C}_{L,s}^{-1}\boldsymbol{C}_{L,q}$, the Eq. (2.55) can be simplified as

$$\boldsymbol{M}\ddot{\boldsymbol{q}}+\left(\boldsymbol{C}_{M,q}+\boldsymbol{C}_{M,s}\boldsymbol{G}\right)^{\mathrm{T}}\boldsymbol{\lambda}_M-\boldsymbol{Q}=\mathbf{0} \tag{2.56}$$

By combining the constraint equations of flexible sliding joint with the dynamic equations, the dynamic equations with the flexible sliding joint can be obtained as

$$\begin{cases}\boldsymbol{M}\ddot{\boldsymbol{q}}+\left(\boldsymbol{C}_{M,q}+\boldsymbol{C}_{M,s}\boldsymbol{G}\right)^{\mathrm{T}}\boldsymbol{\lambda}_M-\boldsymbol{Q}=\mathbf{0} \\ \boldsymbol{C}_M=\mathbf{0}\end{cases} \tag{2.57}$$

in which the internal parameter s needs to be obtained by solving the algebraic equation. Equation (2.57) is a set of nonlinear differential-algebraic equations (DAES), which can be solved by backward difference formulas (BDF).

As shown in Fig. 2.10, an example of moving load along a cantilever beam was used to verify the correctness of the modeling method of the flexible sliding joint. The moving load $\boldsymbol{P}$ acts on the cantilever beam and moves slowly from the left to the right, which can be regarded as a quasi static process. Under the premise of small deformation, the deflection Δ at the right end of the cantilever beam can be given analytically

$$\Delta=\frac{Px^2(3L-x)}{6EI} \tag{2.58}$$

in which EI is the bending stiffness of the beam.

The modeling details of this example are as follows: the cantilever beam is modeled by geometrically accurate beam elements, a slider modeled by rigid body slides along the beam, and the moving load acts on the rigid body. The initial position of the slider is at the left end of the cantilever beam and moves slowly to the right along the beam. The deflection of the right end of the cantilever beam was calculated by the simulation model and the analytical model (normalized by $\Delta/\Delta_{\max}$). Figure 2.11

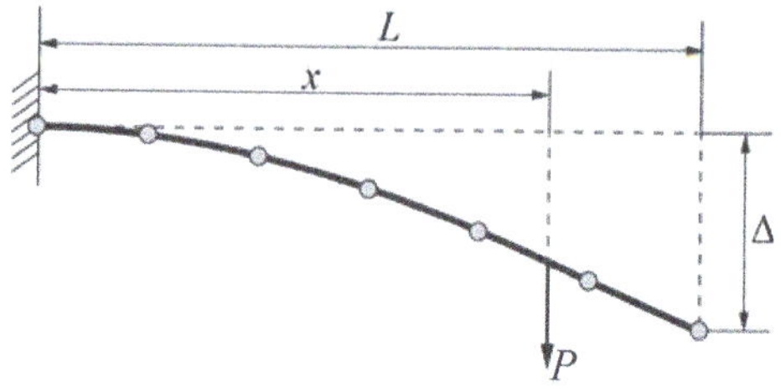

Fig. 2.10 Moving load along a cantilever beam

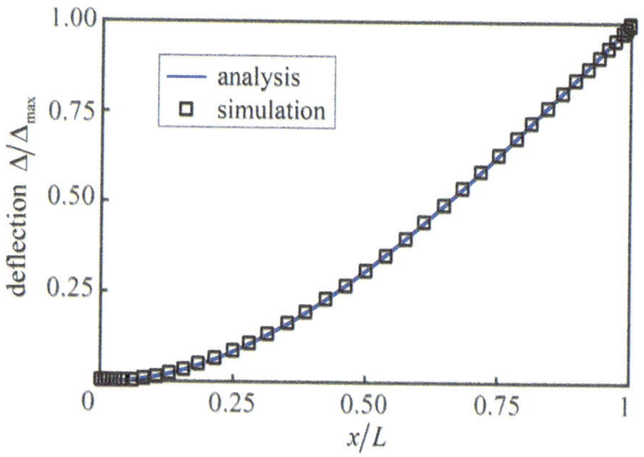

Fig. 2.11 Results of the moving load along a cantilever beam

shows that the two results were completely consistent, indicating the correctness of the modeling method of the flexible sliding joint.

2.4.3 *Stick-Slip Friction of the Sliding Joint*

Considering the friction between sliders and battens in practice, the simulation model should also take into account the influence of friction. For the traditional velocity friction model, the friction coefficient is only a function of the relative velocity, as shown in Fig. 2.12a. The friction force disappears when the relative velocity of the two bodies is zero, which is unreasonable sometimes. For example, if a block is placed on a slope at an angle of θ to the horizontal plane with the static friction coefficient μ_s, the block cannot remain on the slope when the velocity friction model is adopted. Therefore, the velocity friction model cannot describe the viscous state in the friction.

In order to establish a friction model that can describe the viscous state, a stick-slip friction model proposed by Ho-Young Cha [118] is adopted in this section.

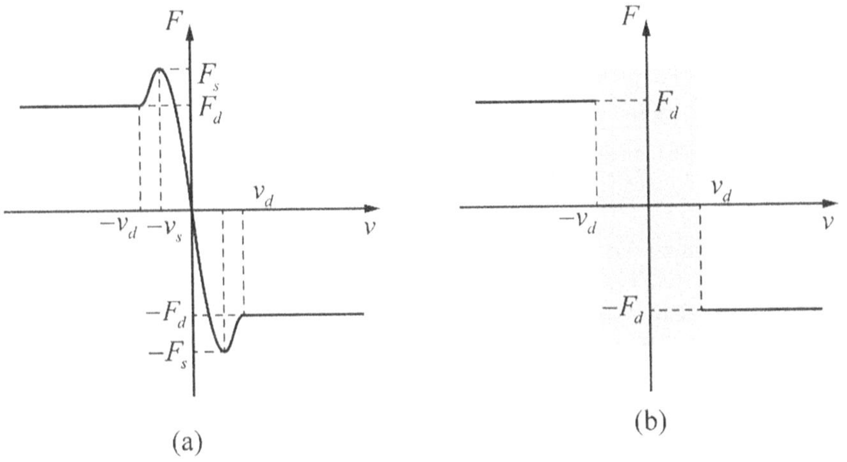

Fig. 2.12 Two types of the friction model: (**a**) velocity friction model; (**b**) Stick-slip friction model

When the relative velocity between the block and the slope is 0, the block can remain static. This stick-slip model can describe the viscous state which is closer to the actual situation. The stick-slip model mentioned above would be introduced into the flexible sliding joint in this section to describe the friction between sliders and battens. The normal contact force f_n can be regarded as a function of embedding depth δ and embedding speed $\dot{\delta}$, and its expression is

$$f_n = -sign(\delta)k|\delta|^m - sign(\dot{\delta})c|\delta|^n \dot{\delta} \tag{2.59}$$

in which k, c, m and n are respectively stiffness, damping coefficient, stiffness index and embedding index, and sign is a sign function.

The key point of stick-slip friction model is that a displacement function is introduced to generate the friction force in the viscous state, and this displacement function is the viscous deformation Δ. The friction force of the dry friction model f is composed of two parts, namely the stick friction force f_{stick} and the sliding friction force f_{slide}:

$$f = f_{stick} + f_{slide} \tag{2.60}$$

The stick friction force f_{stick} and the sliding friction force f_{slide} can be expressed as

$$\begin{aligned} f_{stick} &= -sign(\Delta)(1-\beta)\mu_\Delta(v,\Delta)f_n \\ f_{slide} &= -sign(v)\mu_v(v)f_n \end{aligned} \tag{2.61}$$

where μ_Δ, μ_v, β, and Δ are respectively the viscous friction coefficient, the sliding friction coefficient, the weight of the stick friction, and the viscous deformation. The specific calculation formulations are shown in Table 2.7, in which Δ_{max}, v_t, μ_t

Table 2.7 Calculation formulations of the friction parameters in the stick and sliding state

Parameter	Sliding state $\lvert v\rvert > v_t$	Stick state $0 \le \lvert v\rvert \le v_t$
β	1	STEP($\lvert v\rvert$, $-v_t$, -1, v_t, 1)
μ_Δ	0	STEP($\lvert\Delta\rvert$, $-\Delta_{max}$, $-\mu_t$, Δ_{max}, μ_t)
μ_v	μ_t	STEP($\lvert v\rvert$, $-v_t$, $-\mu_t$, v_t, μ_t)
f	f_{slide}	$f_{slide} + f_{stick}$

are the maximum viscous deformation, the critical velocity and the friction coefficient.

When the relative sliding speed $|v| \le v_t$ and viscous deformation $|\Delta| > \Delta_{max}$, the stick friction force f_{stick} and sliding friction force f_{slide} are respectively as follows

$$\begin{aligned} |f_{stick}| &= \left[1 - \mathrm{STEP}\left(|v|, -v_t, -1, v_t, 1\right)\right]\mu_t f_n \\ |f_{slide}| &= \mathrm{STEP}\left(|v|, -v_t, -1, v_t, 1\right)\mu_t f_n \end{aligned} \tag{2.62}$$

Hence the total friction force

$$|f| = \mu_t f_n \tag{2.63}$$

When the relative sliding speed $|v| \le v_t$ and viscous deformation $|\Delta| \le \Delta_{max}$, the stick friction force f_{stick} and sliding friction force f_{slide} are respectively as follows

$$\begin{aligned} |f_{stick}| &= \left[1 - \mathrm{STEP}\left(|v|, -v_t, -1, v_t, 1\right)\right]\mathrm{STEP}\left(|\Delta|, -\Delta_{max}, -\mu_t, \Delta_{max}, \mu_t\right) f_n \\ |f_{slide}| &= \mathrm{STEP}\left(|v|, -v_t, -1, v_t, 1\right)\mu_t f_n \end{aligned} \tag{2.64}$$

Hence the total friction force

$$|f| < \mu_t f_n \tag{2.65}$$

Figure 2.12b shows the relationship between the friction coefficient and the relative velocity when the stick deformation is $\Delta = \Delta_{max}$. It can be seen that the stick friction force still exist when the relative velocity is 0, hence this model can describe the viscous state.

A simple example is demonstrated to verify the correctness of stick-slip friction modeling method. As shown in Fig. 2.13, a slider with a mass of $m = 1$ kg slides along the simply supported beam. The slider is driven by the sinusoidal force $F = 1.5 \sin t$ N along the beam, the friction coefficient between the slider and the beam $\mu = 0.1$, and the gravitational acceleration $g = 10$ m/s^2. According to the force analysis, the maximum static friction force $f_{max} = 1$ N. When the driving force F is larger than 1 N, relative sliding will occur between the slider and the beam. When the driving force F is less than 1 N, the slider will remain static relative to the beam, that is, the slider will remain in a viscous state.

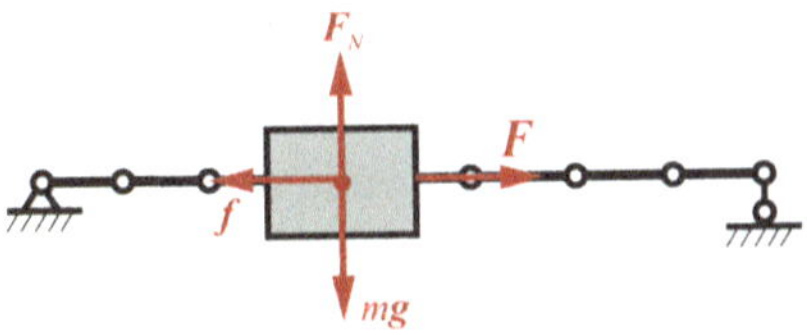

Fig. 2.13 A slider along a simply supported beam

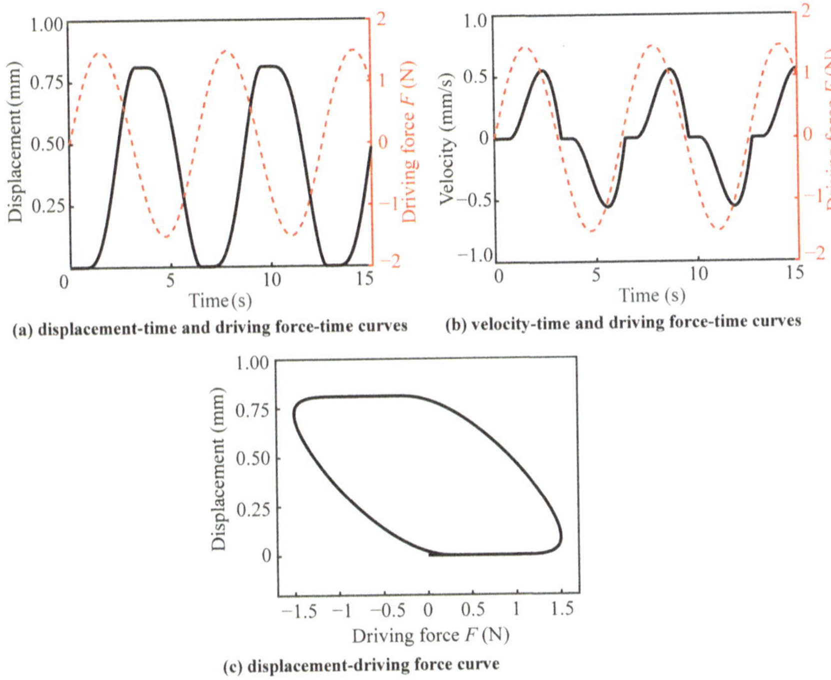

Fig. 2.14 Simulation results of the stick-slip friction model: (**a**) displacement-time and driving force-time curves; (**b**) velocity-time and driving force-time curves; (**c**) displacement-driving force curve

The slider and simply supported beam were modeled with rigid body and geometrically accurate beam respectively, the flexible sliding joint between the slider and beam was established, the sinusoidal force was applied to the slider, and the stick-slip friction was applied between the slider and beam. Figure 2.14a and b respectively show the curves of displacement, velocity and force with time. It can be seen that when the driving force F is less than 1 N, the slider remains static. Figure 2.14c shows the relationship between the displacement of the slider and the driving force. It can be seen that the displacement produces "hysteresis" with the driving force, indicating that the slider remains static when the driving force is less than the maximum static friction force.

2.4.4 *Assembly and Calculation of the Governing Equations*

Eventually, the governing equations of the mesh antenna form a set of differential algebraic equations (DAEs)

$$\begin{cases} \boldsymbol{M}\ddot{\boldsymbol{q}} + \left(\dfrac{\partial \boldsymbol{C}}{\partial \boldsymbol{q}}\right)^{\mathrm{T}} \boldsymbol{\lambda} - \boldsymbol{Q}(\boldsymbol{q},\dot{\boldsymbol{q}},t) = \boldsymbol{0} \\ \boldsymbol{C}(\boldsymbol{q},\dot{\boldsymbol{q}},t) = \boldsymbol{0} \end{cases} \tag{2.66}$$

where $\boldsymbol{q}$ is the generalized coordinate vector of the system, $\boldsymbol{M}$ is the assembled mass matrix, and $\boldsymbol{C}(\boldsymbol{q},\dot{\boldsymbol{q}},t) = \boldsymbol{0}$ represents the assembled constraint equations. In particular, it is worth mentioning that the corresponding part of the Jacobian matrix of constraints $\partial \boldsymbol{C}/\partial \boldsymbol{q}$ should be replaced with $\partial \boldsymbol{C}_s/\partial \boldsymbol{q}_s - \partial \boldsymbol{C}_s/\partial \boldsymbol{\varphi}(\partial \boldsymbol{C}_t/\partial \boldsymbol{\varphi})^{-1}\partial \boldsymbol{C}_t/\partial \boldsymbol{q}_s$, when the synchronization cable-pulley mechanisms are assembled into the whole system. The DAEs can be directly solved by stiff numerical integrators such as backward differentiation formulas (BDF). Similar to the two-bay model, the entire model of the mesh antenna is built and simulated based on our lab's in-house code.

2.5 Design and Verification of Synchronization Mechanism

In Astromesh-II, the translational joints of two adjacent bays are synchronized via cable-pulley mechanisms. In this section, we designed two different synchronization mechanisms and compared their effects on deployment dynamics of the mesh antenna.

2.5.1 *Variant-Wrap-Angle Design*

As shown in Fig. 2.15, there are two translational joints in a single bay, and the original degree of freedom is two. However, if two cables across pulleys are used to connect the two sliders, when slider 1 moves down, cable 1 (solid line) will pull the slider 2 up; when the slider 1 moves up, cable 2 (dashed line) pulls the slider 2 down. The movement of the slider 1 restricts the movement of the slider 2, and vice versa. The ideal aim is to achieve the system with a single degree of freedom, in which the cable connecting the slider makes the two adjacent sliders with the same displacement, that is, $\Delta_1 = \Delta_2$. Therefore, cables 1 and 2 are also called synchronization cables, and this mechanism is also called synchronization cable-pulley mechanism.

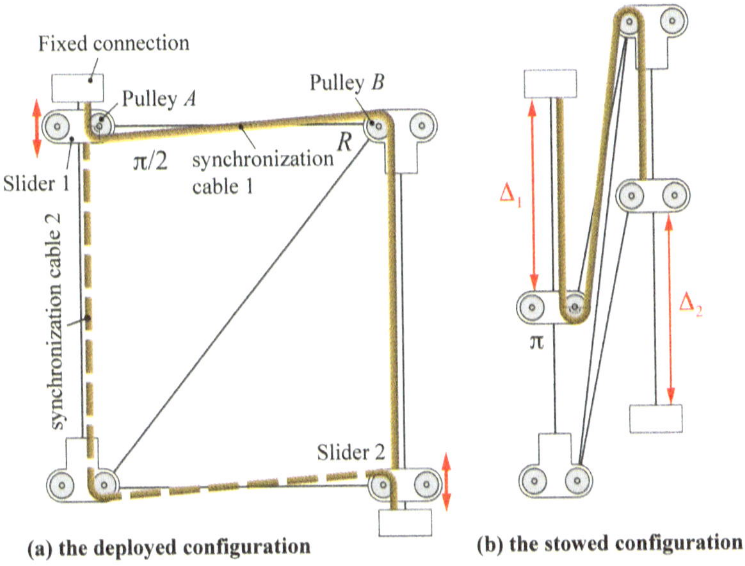

Fig. 2.15 Variant-wrap-angle design: (**a**) the deployed configuration; (**b**) the stowed configuration

The synchronization mechanism adopts the variant-wrap-angle design, as shown in Fig. 2.15. Taking the synchronization cable 1 as an example, one end of the cable is fixed on the battens at the top left. The synchronization cable 1 first goes down parallel to the left batten, then bypasses the pulley *A* and turns to the transverse right, and then bypasses the pulley *B* again and turns down vertically to the slider 2. The radius of the two pulleys is $R = 7$ mm, far less than the length of the longerons $L_2 = 530.2$ mm. Therefore, the synchronization cable between pulley *A* and pulley *B* can be approximately considered to be parallel to the longeron and equal in length to the longeron.

If the moving distances of the two sliders are the same, the sum of the vertical and parallel lengths of the two segments of the synchronization cable remains a constant during the movement geometrically. However, the length of the cable wrapped around the pulley has changed. In the deployed state shown in Fig. 2.15a, the wrapping angle of the cable on each pulley is approximately $\pi/2$, and the length of the cable wrapped on the two pulleys approximately equals to πR. In the stowed state shown in Fig. 2.15b, the wrapping angle of the cable on each pulley is approximately π, and the length of the cable wrapped on the two pulleys approximately equals to $2\pi R$. Therefore, this design is called Variant-wrap-angle design in this chapter. That is to say, the increase of the length of the synchronization cable from the deployed configuration to the stowed configuration approximately equals to

$$\Delta L \approx R\pi = 17.9\,\text{mm} \tag{2.67}$$

If the synchronization cable is tensioned and approximately unstretchable, the increase of the length of the synchronization cable in the folding process means that the freedom of the single span is zero and it cannot be folded.

In practice, the synchronization cable is extensible. Considering the extension of the synchronization cable, for the stowed configuration, when the strain of the cable reaches

$$\varepsilon = \Delta L / L_0 = 9.0018 \times 10^{-3} \tag{2.68}$$

The corresponding tension of the cable is

$$T = EA\varepsilon = 1.7675 \times 10^3 \text{ N} \tag{2.69}$$

It can be seen that the tension of the synchronization cable is nearly 2000 N, which is a significant force that the truss cannot sustain.

If the synchronization cable is slightly tensioned (with the tensioning force 20 N) in the stowed state, the tension value of the synchronization cable gradually decreases during the unfolding process and the synchronization cable is slack in the unwound state. As a result, the synchronization cable cannot guarantee the same displacement of the two adjacent sliders. The difference of displacement is more than 10 mm, which can be accumulated after multiple bays and greatly weaken the synchronization effect. Therefore, the variant-wrap-angle design cannot achieve good performance of synchronization, and the synchronization mechanism should be redesigned.

2.5.2 *Constant-Wrap-Angle Design*

Figure 2.16 is the schematic diagram of the constant-wrap-angle design. Taking the synchronization cable 2 as an example, one end of the synchronization cable 2 is fixed on the slider 1. The synchronization cable 2 first goes down parallel to the left batten, then bypasses the pulley *A* and then turns to the right parallel to the longeron, and then bypasses the pulleys *B* and *C* on the sliders in sequence. Finally, the cable goes downwards parallel to the right batten and the other end is fixed on the batten. This design has the following points for attention:

1. The synchronization cable passes through the pulley for three times, and the pulley *B* and pulley *C* are fixed to the same slider;
2. The radius of pulley *A* is equal to that of the pulley *B*, and the axes of rotation of the pulley *A* and the pulley *B* are also required to be the axes of rotation of the longeron;
3. For the deployed configuration, the wrapping angle of the synchronization cable on the pulley *B* should be no less than 90 degrees.

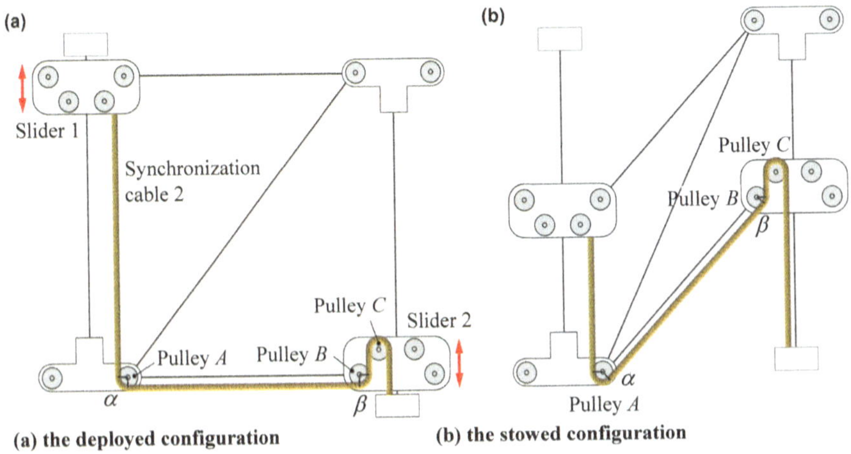

Fig. 2.16 Constant-wrap-angle design: (**a**) the deployed configuration; (**b**) the stowed configuration

Simple geometric analysis shows that the sum of the three lengths of the synchronization cable parallel to the longerons or battens remains unchanged in the unfolding process. The following focuses on the length of the synchronization cable wrapped around the pulleys. For the deployed configuration shown in Fig. 2.16a, the wrapping angle on pulley A and pulley B is $\pi/2$, the wrapping angle on pulley C is π, and the cable length wrapped on the three pulleys is $2\pi R$. From the deployed to stowed configuration shown in Fig. 2.16b, the wrapping angle on pulley A increases, the wrapping angle on pulley B decreases, and the wrapping angle on pulley C remains the same. Geometrically, it can be easily proven that the sum of wrapping angles on the three pulleys always remains to be π, i.e., $\alpha + \beta = \pi$, resulting in the cable length wrapped on the pulleys keeps constant. To sum up, the total length of the synchronization cable is constant in the unfolding process. This design is also called constant-wrap-angle design.

Constant-wrap-angle design geometrically ensures that the length of each synchronization cable is constant during the unfolding process. Theoretically, if the synchronization cable is tensioned when the antenna is stowed, the synchronization cable will keep taut throughout the unfolding process. Therefore, the constant-wrap-angle design is expected to achieve good performance in synchronization.

2.5.3 *Experimental Validation of the Constant-Wrap-Angle Design*

In order to further verify the synchronization effect of the proposed constant-wrap-angle design, experimental validation was carried out on the lab's in-house six-bay ring truss with 2-m diameter as shown in Fig. 2.17a. The parameters are shown in

(a) Overview of the model, the number of the translational joint starts from 0 at the far end of the top view and increase counterclockwise;

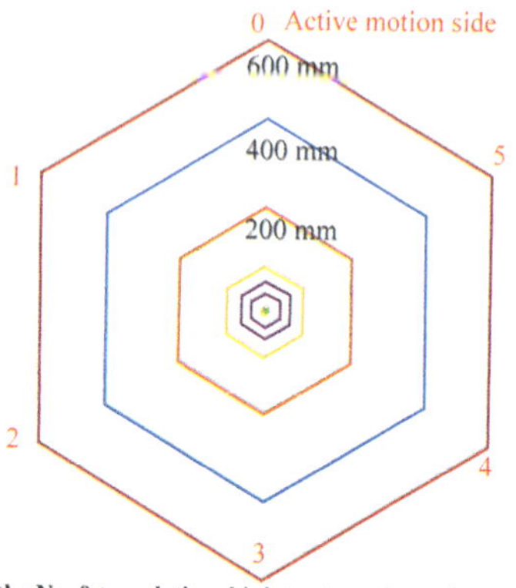

(b) the No. 0 translational joint acts as the active motion, and the others move passively

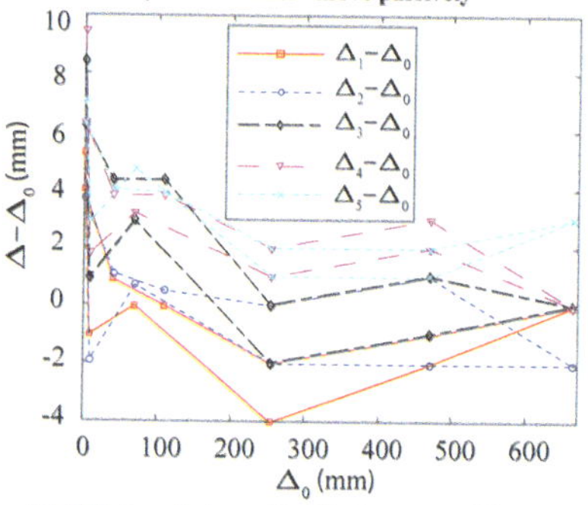

(c) difference between the displacement of the translational joint and the active motion

Fig. 2.17 The synchronization experiment of the six-bay ring truss with 2 m diameter: (**a**) overview of the model; (**b**) the No. 0 translational joint acts as the active motion, and the others move passively; (**c**) difference between the displacement of the translational joint and the active motion

Table 2.8 Parameters of the six-bay ring truss with 2-m diameter

Parameter	Value	Parameter	Value
Diameter	2 m	Height	1.4 m
Number of bays	6	Elastic modulus of the truss	100 GPa
Outer diameter of the truss	25.2 mm	Inner diameter of the truss	24 mm

Table 2.8. As shown in Fig. 2.17b, from the top view, the number of translational joints is 0 at the far end and increases counterclockwise in turn. The No. 0 translational joint acts as the active motion, and the other translational joints (from No. 1 to No. 5) move passively under the action of the synchronization cable. Specifically, the displacement of the No. 0 translational joint was set to be $\Delta_0 = 0, 7.9, 68.6, 254, 470, 663$ mm respectively. If the synchronization cable ensures good synchronization, the other five translational joints will keep the same movement with the No. 0 joint, resulting in the transformation from the deployed configuration to the stowed configuration. Then, the opposite displacement driving $\Delta_0 = 663, 470, 254, 109.2,$

Table 2.9 The synchronization experiment of a six-bay ring truss with 2 m diameter

	Measuring implement	Translational displacement (mm)					
		Δ_0 (active motion)	Δ_1	Δ_2	Δ_3	Δ_4	Δ_5
Folding process	Vernier caliper	0.0	5.3	6.3	8.4	9.5	7.1
		7.9	6.9	6.0	8.8	9.7	10.8
		68.6	68.6	69.3	71.5	71.8	73.3
	Tapeline	254	250	252	252	255	255
		470	468	468	469	472	471
		663	663	661	663	663	666
Unfolding process	Tapeline	470	469	471	471	473	472
		254	252	254	254	256	256
	Vernier caliper	109.2	109.2	109.7	113.5	113.0	113.1
		39.4	40.3	40.5	43.7	43.2	43.4
		0.0	4.0	3.7	6.2	6.3	6.3

39.4, 0 mm was applied to the No. 0 translational joint, resulting in the transformation from the transformation from the deployed configuration to the stowed configuration to the deployed configuration. In the above experiment, the motion of the No. 0 translational joint was applied manually.

In the experiment, the displacements of five translational joints Δ_n, $n = 1, 2, \cdots, 5$ in the folding and unfolding process were measured. When the displacement of the translational joints is less than 150 mm, it was measured with a vernier caliper (measurement error ± 1 mm). When the displacement of the translational joints is more than 150 mm, it was measured with a tapeline (measurement error ± 2 mm). The measured data were shown in Table 2.9, which presents good synchronization. In order to compare the differences more intuitively, the differences between the displacements of translational joints and the active motion were further obtained, that is, $\Delta_n - \Delta_0$, $n = 1, 2, \cdots, 5$. As shown in Fig. 2.17c, the largest displacement difference of each bay occurs in the deployed state, which was about 10 mm. The maximum difference of the transition state during unfolding and folding process was about 6 mm.

Therefore, the synchronization mechanism with the constant-wrap-angle design ensures that the length of the synchronization cable does not change during the unfolding process, and the displacement of each translational joint of the ring truss is nearly the same. Therefore, it can be concluded that the constant-wrap-angle design has the advantages of simple structure and good synchronization.

2.5.4 *Material and Pretension of the Synchronization Cable*

For the design of the synchronization cable-pulley system, it is necessary to consider the material and initial tension of the synchronization cable. At the beginning, the fishing line with a diameter of 0.47 mm was used for the synchronization cable.

Table 2.10 Creep deformations of cables

Time	Creep deformation (mm)	
	Steel wire	Fishing line
22 min	0	0
40 min	0	7
1 h and 40 min	0	10
3 h and 40 min	0	15
1 day	0	19
2 days	0	28
3 days	0	36

However, it was found that there existed creep deformation on the fishing line, resulting in the initial tension disappeared after a few days. Later on, the steel wire with a diameter of 0.45 mm was used for the synchronization cable. The creep deformations of both the steel wire and fishing line were measured quantitatively. The method was to make two 2 m cables with steel wire and fishing line with a weight of 4 kg suspended respectively. The creep deformations measured with a tapeline are shown in Table 2.10, and the measurement error was about ±1 mm. Over a period of 3 days, the creep deformation of the steel wire was approximately zero, while the fishing line stretched 36 mm. According to the cognition of the synchronization cable, it is better to choose the materials with large tensile modulus, small bending modulus and no creep deformation. At present, the steel wire is used in the experiment of the six-bay ring truss with 2 m diameter, and the synchronization effect is good.

Another concern with the synchronization cable is to set the pretension with the appropriate value. A large tension can make up for some gaps in the mechanical design, such as the 0.5 mm gap between pulleys and shafts of the six-bay ring truss with 2 m diameter. However, the initial tension will lead to the increase of the friction between pulleys and the rotating shafts, resulting in large driving force. Finally, a smaller pretension of the synchronization cable, such as 30 N, was used in the experiment of the six-bay ring truss with 2 m diameter.

2.6 Evaluation of Synchronization Mechanisms on Deployment Dynamics

Two numerical models of the 30-bay ring truss and the cable networks were tested, as shown in Fig. 2.7. The only difference between these two models is the synchronization mechanism, where one uses the variant-wrap-angle design and the other one uses the constant-wrap-angle design. In the stowed configuration, the synchronization cables in both models are pre-tensioned by a small tensile force, 30 N, to compensate for the joint clearances in manufacture and to ensure that all of the synchronization cables are taut.

Here the synchronization performance of the two designs is evaluated and compared through three aspects: kinematic latch time, bending moments of the ring truss, and the driving forces.

2.6.1 Kinematic Latch Time

As shown in Fig. 2.18a, the simulation results for the variant-wrap-angle design show that the ring truss unfolds asynchronously—there is remarkable displacement delay in the translational joints from the near-motor side to the far-motor side in sequence. Specifically, the five bays near the motors begin to latch into the fully deployed configuration when $t = 565$ s, then the remaining 25 bays are sequentially locked, as shown in Fig. 2.18b. The duration of latching time from the first bay to the last bay lasts for 160.5 s. This is caused by the fact that the synchronization cables are taut at the stowed configuration while slack at the deployed configuration.

In contrast, for the constant-wrap-angle design, the simulation results show approximately synchronous deployment. All of the bays unfold with almost the same speed. The time difference between the latching of the last bay and the latching of the first bay is 660.8–644.9 = 15.9 s, only 1/10 the time difference of that in variant-wrap-angle design. In addition, the tensions of the synchronization cables during the unfolding process are always positive, or taut, actively ensuring synchronization.

In addition, Fig. 2.19 shows the tension changes of the synchronization cables with two synchronization mechanisms during the whole unfolding process. It can be seen that the synchronization cables with constant-wrap-angle design are tensioned in the unfolding process. However, the synchronization cables with variant-wrap-angle design may be slack in the unfolding process. The constant-wrap-angle design ensures that the synchronization cable is tensioned in the unfolding process, and the synchronization cable is effective in the whole process of unwrapping, resulting in good synchronization of the deployable ring truss. For the variant-wrap-angle design, the slack synchronization cable cannot guarantee the synchronization effect, and the synchronization effect is greatly weakened, resulting in the obvious asynchronization.

2.6.2 Bending Moments of the Ring Truss

As expected in the introduction, simulations confirmed that the synchronization performance is closely related to the bending moments of the ring truss. Figure 2.20 shows the historical maximum bending moment of each longeron and batten during the entire unfolding process.

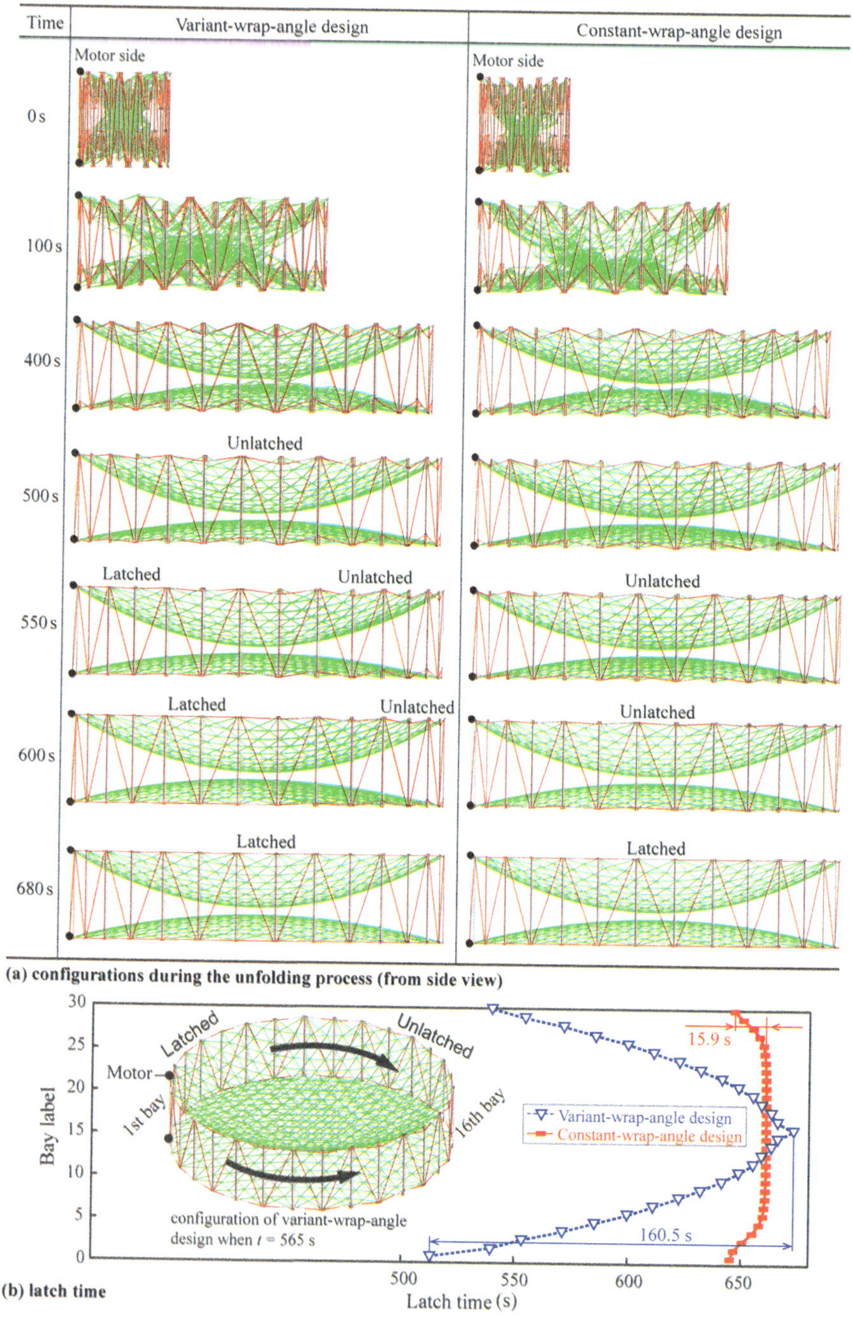

Fig. 2.18 Configurations during unfolding process and latch time: (**a**) configurations during the unfolding process (from side view); (**b**) latch time. Adapted with permission from [107]

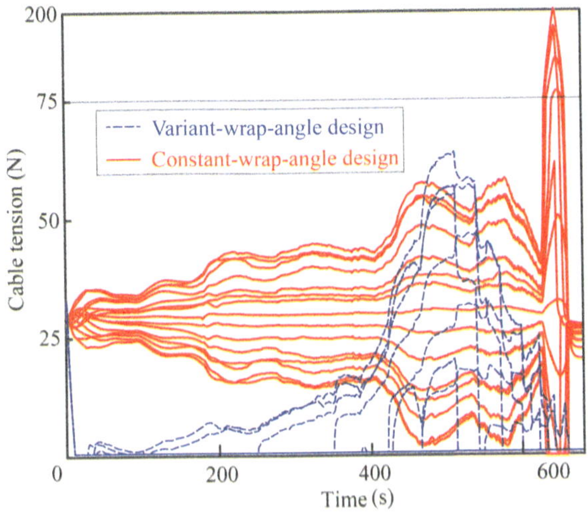

Fig. 2.19 The tension variations of two synchronization mechanisms in unfolding process. Adapted with permission from [107]

For longerons, there is no load applied between their two rotation jointed ends, resulting in little bending moments in the parallelogram plane of each bay [109]. The simulation results also proved this, as the maximum moment of the longerons in the parallelogram plane was less than 0.002 N·m for both the variant-wrap-angle design and constant-wrap-angle design. However, the bending moments in the cross plane of the cylindrical ring truss differ. As shown in Fig. 2.20a, the maximum bending moment in the cylindrical cross plane reaches nearly 10.7 N·m in the variant-wrap-angle model, over six times the bending moment in the constant-wrap-angle model, which was 1.4 N·m.

For battens, the bending moments in both the radial and tangent planes of the cylindrical ring truss were measured. The maximum bending moments of the two planes are plotted in Fig. 2.20b respectively for the two designs. Compared with the peak moment of 5.3 N·m for the variant-wrap-angle design, the maximum moments for the constant-wrap-angle design reached 9.6 N·m. The larger moments in the constant-warp-angle model may be caused mainly by the significant tensions of the synchronization cables, which reached over 200 N when the ring truss was nearly fully deployed, as shown in Fig. 2.19, while most of the synchronization cables in the variant-wrap-angle model were slack at this stage of deployment.

2.6.3 *Potential Energy and Butterfly Effect on the Driving Force*

Figure 2.21a and b plot the time-histories of the potential energy curves of the cable networks and the driving force of one motor during the unfolding process for each of the two designs. The varying trend of the two forces and the two energy curves is same—they remain low for a long period of time and then sharply increase as the antenna approaches complete deployment. This is caused by the tension in the cable

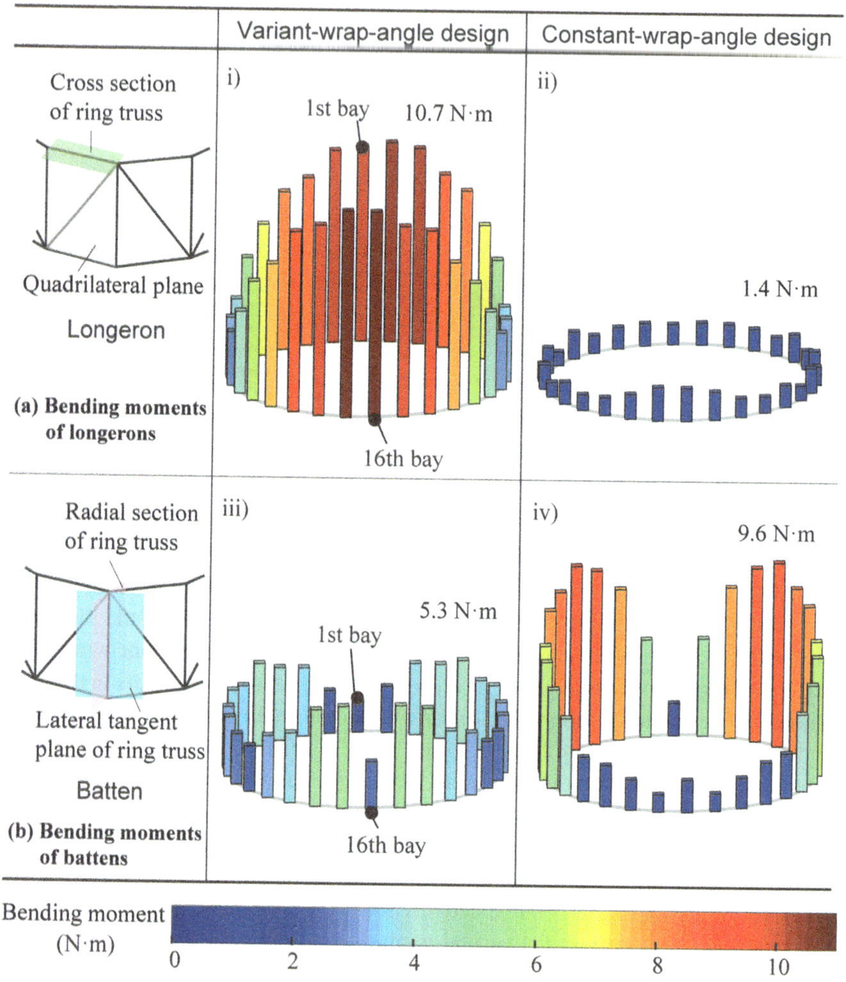

Fig. 2.20 The tension variations of two synchronization mechanisms in unfolding process: (**a**) bending moments of longerons; (**b**) bending moments of battens. Adapted with permission from [107]

networks at nearly full deployment. Figure 2.21c and d show the variation curve of the antenna potential energy during the unfolding process. At the initial stage, the potential energy is very small due to no tension of the sky cable networks, while when the antenna is nearly fully deployed, the potential energy increases sharply due to the tension of the cable networks. Considering that the increase of antenna potential energy comes from the work done by motor, it is reasonable that the driving force of the motor is consistent with the variation trend of antenna potential energy.

During the unfolding process, part of the input energy of motors is stored as potential energy of the ring truss and cable networks, and part is dissipated by friction. Friction dissipation mainly comes from the friction of the translational joints

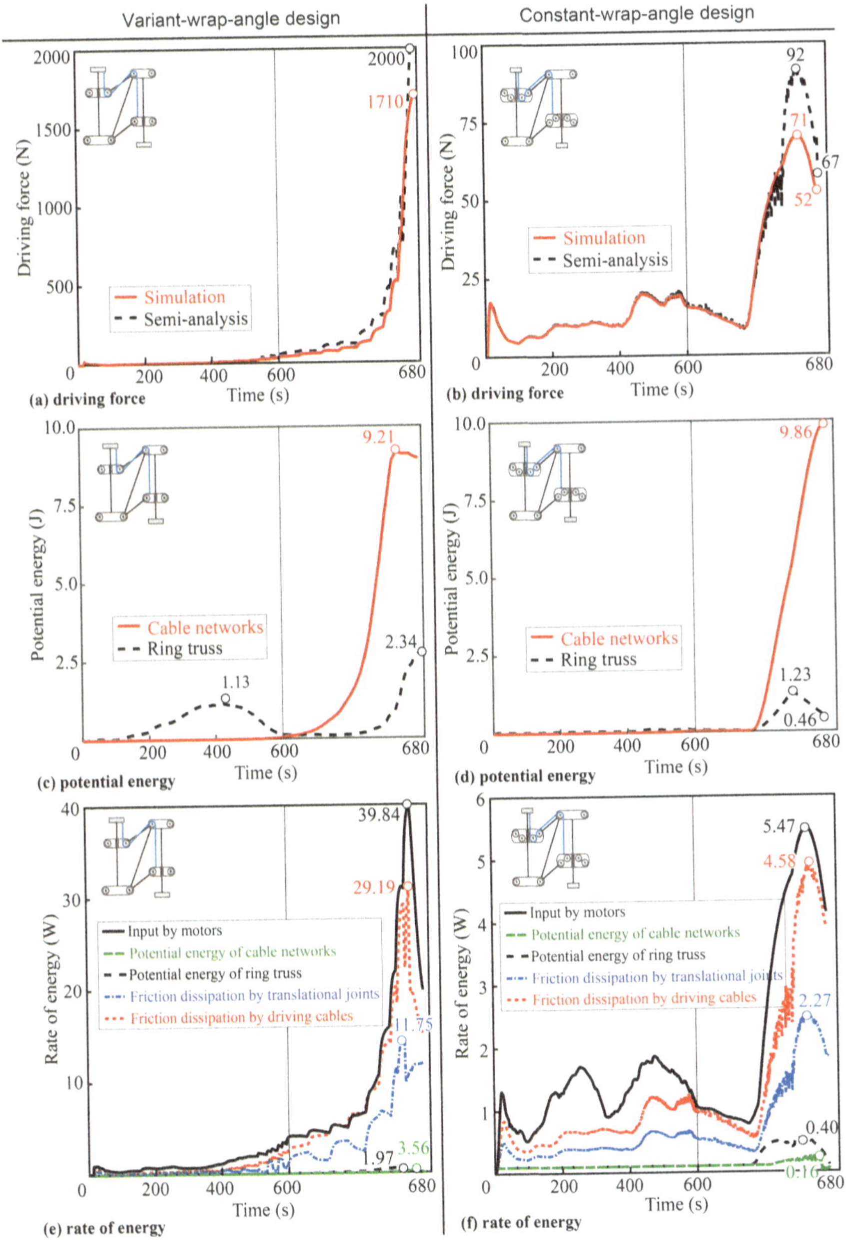

Fig. 2.21 Energy analysis of the variant-wrap-angle design and constant-wrap-angle design: (**a**, **b**) driving force; (**c**, **d**) potential energy; (**e**, **f**) rate of energy. Adapted with permission from [107]

along battens and the tension decay of the driving cables passing through pulleys. Therefore, from the energy point of view, the following relationship can be established as

$$4Tv_{\text{motor}} = \dot{E}_{\text{web}} + \dot{E}_{\text{truss}} - \dot{W}_{\text{trans}} - \dot{W}_{\text{dri}} \tag{2.70}$$

in which $4Tv_{\text{motor}}$ is the total power input by the motors, 4 represents the number of motors, T is the driving force of the motor, and v_{motor} is the reel-in velocity of the motor. $\dot{E}_{\text{web}}$ and $\dot{E}_{\text{truss}}$ denote the change rate of potential energy of cable networks and ring truss with time respectively; $\dot{W}_{\text{trans}}$ and $\dot{W}_{\text{dri}}$ represent the friction dissipation rate of the translational joints between sliders and battens and the driving cables through pulleys respectively. The friction dissipation rate of the translational joints and the driving cables can be expressed as

$$\begin{aligned}\dot{W}_{\text{trans}} &= \sum_{i=1}^{30} f_{\text{trans}}^{i} v_{\text{trans}}^{i} \\ \dot{W}_{\text{dri}} &= \sum_{i=1}^{n_{\text{d}}} (1-\eta) T_{\text{syn}}^{i} v_{\text{syn}}^{i}\end{aligned} \tag{2.71}$$

in which f_{trans}^{i} and v_{trans}^{i} represent the friction force of translational joints and the relative velocity between sliders and battens; T_{syn}^{i} and v_{syn}^{i} represent the tension of the driving cable at the i-th segment and the relative velocity between the driving cable and the pulley; n_{d} is the number of the pulleys which keep in contact with driving cables. The variation rates of work done, potential energy and friction dissipation of the motors with time are shown in Fig. 2.21e and f. The friction dissipated energy is much larger than the potential energy stored by the antenna, hence most of the input energy of the motors is dissipated by friction.

Substituting Eq. (2.71) into Eq. (2.70), a semi-analytical expression of the motor driving force T can be obtained, as shown by the dotted black curve in Fig. 2.21a and b. It can be seen that the semi-analytical results of driving force are in good agreement with the simulation results, which further proves that the above energy analysis is reasonable.

As can be seen from Fig. 2.21a and b, although two different synchronization mechanisms are adopted, the variation trend of driving forces is similar. The amplitude of driving forces is greatly different. The maximum driving force of the variant-wrap-angle design reaches 1700 N, while the maximum driving force of the constant-wrap-angle design is less than 100 N. The reasons for the large difference in the driving force peak value are discussed below.

This might be caused by a butterfly effect in the driving mechanism. As shown in Fig. 2.22a, near the end of deployment in the variant-wrap-angle design, the asynchronous behavior results in only one or a few translational joints at the far-motor side transferring energy into the cable networks, so the required tension of the driving cable at the far-motor bay is large. It then needs much more tension at the near-motor side because the friction decay at the far motor side is significant. However, in the constant-wrap-angle design shown in Fig. 2.22b, the synchronous behavior allows many translational joints, both at the far-motor side and the near-motor side, to transfer energy into the cable network. Therefore, the tension in the unfolding bays is relatively small, which in turn requires a smaller amount of tension at the near-motor side.

The simulation results above show that the synchronization mechanism plays an important role in the deployment of the mesh antenna. As shown in Fig. 2.22c,

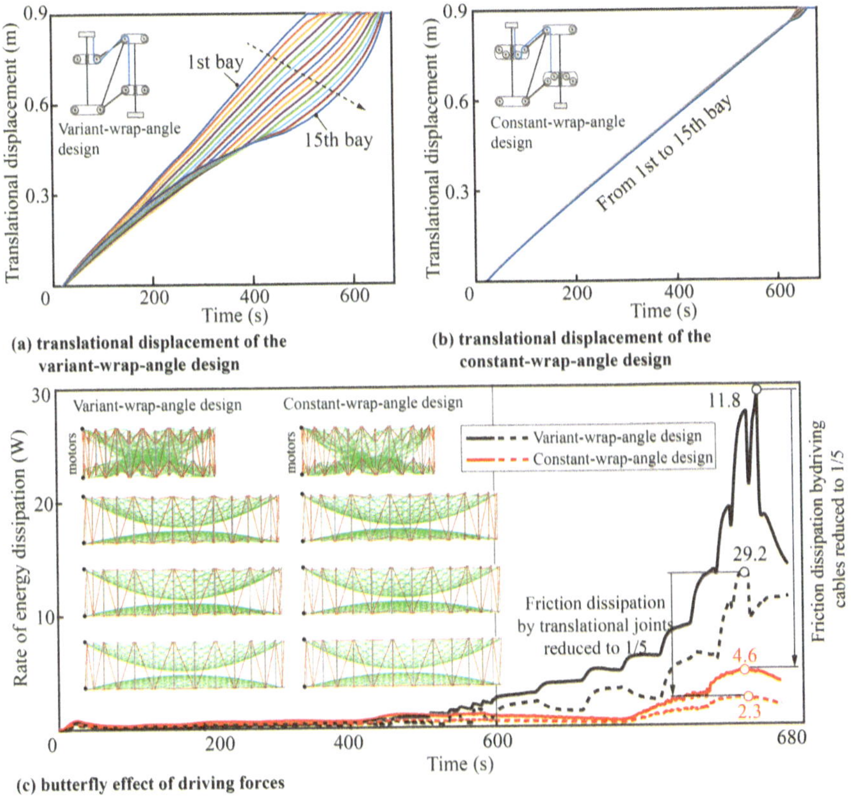

Fig. 2.22 Translational displacements and butterfly effect of driving forces: (**a**) translational displacement of the variant-wrap-angle disign; (**b**) translational displacement of the constant-wrap-angle design; (**c**) butterfly effect of driving forces

compared with the variant-wrap-angle design, the constant-wrap-angle design contributes to better synchronous deployment, which in turn requires a much smaller driving force from the motors, further resulting in smaller bending moments in the ring truss. Therefore, it is straightforward to conclude that the constant-wrap-angle design is better for antenna deployment, and that design is also adopted in the following further studies.

2.6.4 *Overall Comparison of the Two Synchronization Mechanisms*

Two different synchronization mechanisms of the variant-wrap-angle design and the constant-wrap-angle design are shown in Fig. 2.23a. The influence of the two synchronization mechanisms on antenna deployment dynamics is comprehensively compared as shown in Fig. 2.23b–d, including the latch time, the bending moment and the driving force of motor. Compared with the variant-wrap-angle design, the

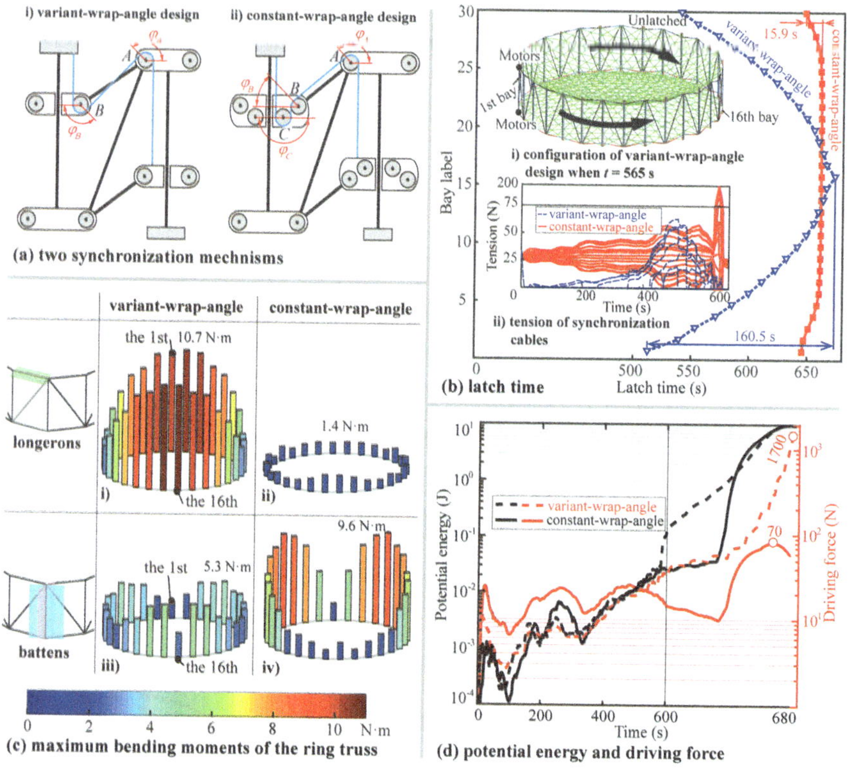

Fig. 2.23 Design of synchronization mechanisms and influence on deployment dynamics: (**a**) two synchronization mechnisms; (**b**) latch time; (**c**) maximum bending moments of the ring truss; (**d**) potential energy and driving force. Adapted with permission from [107]

constant-wrap-angle design contributes to better synchronous deployment, which can ensure good synchronization in the unfolding process, and synchronous unfolding is helpful to reduce the bending moment of the ring truss and the maximum driving force of the motors. In order to ensure the deployment reliability of antennas in orbit, the design which can ensure synchronous deployment is welcomed, such as the constant-wrap-angle design.

2.7 Friction Sensitivity Analyses

For the constant-wrap-angle design, numerous simulations were conducted to investigate the friction sensitivity on the deployment. The system contains two main kinds of friction: friction within the driving cables, described by the transmission efficiency η, and friction within the translational joints, described by the translational friction coefficient μ. The friction force within a translational joint was calculated by multiplying μ and the constraint force of the joint. Here, $5 \times 5 = 25$ models with different friction coefficients were simulated, in which the transmission efficiency $\eta \in [0.84,\ 1]$ with intervals of 0.04, and the translational friction coefficient $\mu \in [0,\ 0.20]$ with intervals of 0.05. The historical maxima of two variables were monitored during deployment: peak bending moment M_{bend} of all of the longerons and battens, and peak driving force T_{driving}. These are functions of the friction coefficients, which can be expressed as

$$\left(M_{\text{bend}}, T_{\text{driving}}\right) = \mathbf{g}\left(\eta, \mu\right) \tag{2.72}$$

The results are plotted in Fig. 2.24, whose surfaces were fitted based on the 25 individual points by using a biharmonic spline interpolation called griddata method.

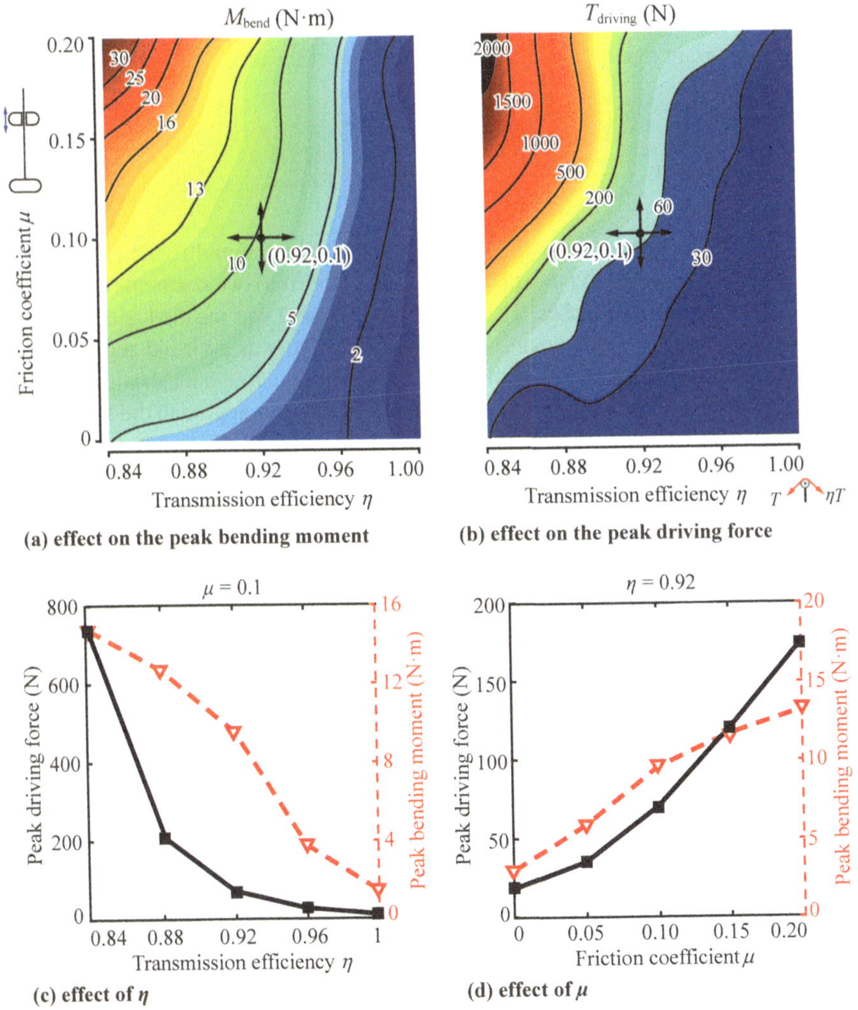

Fig. 2.24 Sensitivity of η and μ on the deployment: (**a**) effect on the peak bending moment; (**b**) effect on the peak driving force; (**c**) effect of h; (**d**) effect of μ. Adapted with permission from [107]

Figure 2.24a and b shows that M_{bend} and T_{driving} share a common variation trend: they are enlarged by decreasing η or increasing μ. In an extreme case without friction, or $(\eta, \mu) = (1.00, 0)$, the driving force as small as 5.8 N is enough to successfully unfold the antenna with invisible deformation of the ring truss. However, in the other extreme case, when the two friction coefficients reach their border maxima, or $(\eta, \mu) = (0.80, 0.20)$, the motors fail to deploy the mesh antenna, as emphasized by the black area in Fig. 2.24b. The reason is that the driving force reaches 2000 N, and the reel-in velocity thus decreases to zero.

Further qualitative analysis in Fig. 2.24c and d reveals that the driving force is more sensitive to η than to μ, as variation of η causes nearly three orders of magnitude difference in the driving force, while variation of μ causes about two orders of magnitude difference. Moreover, η and μ have comparable effects on the bending moment; they both result in a small change of ~10 N·m. Overall, the effect of friction on the driving force is much more notable than on the bending moment, and η is more important than μ.

2.8 Summary

This chapter established a novel methodology to model the multiscale physics of a cable-pulley system. We dynamically divided the cable into two segments and treated them with different approaches. The noncontact segment was meshed with ALE elements, while the contact segment was considered merely as a virtual segment constraining the border of the ALE element. With this method, rough mesh was sufficient to take the relevant multiscale physics into account, and this approach did not involve any detection of contact events. Therefore, the calculation efficiency was significantly improved, which was validated by a two-bay model of the Astromesh-II.

The proposed multiscale method was then used to evaluate and design the synchronization mechanism in Astromesh-II. In terms of deployment dynamics, two different designs were compared for the synchronization mechanisms: a variant-wrap-angle design and a constant-wrap angle design; the friction of the cable-pulley systems and the translational joints and its effects were also investigated. Results show that the variant-wrap-angle design leads to remarkable asynchronous deployment, while the constant-wrap-angle design contributes to nearly synchronous deployment, accompanied by smaller bending moments in the ring truss and smaller driving forces by motors. Therefore, the constant-wrap-angle design is recommended in practice. For the constant-wrap-angle design, both the peak bending moment and the peak driving force rely on the transmission efficiency of the driving cables and the translational friction. Further sensitivity analyses revealed that the driving force is more sensitive to the transmission efficiency than the translational friction. However, the two kinds of frictions have comparable effects on the bending moment. These insights are ready to be applied in actual scenarios, and the developed multiscale modeling method can be used to study and understand other similar mechanical mechanisms.

Chapter 3
Robust Design of Cable Networks Against Variation in Tension Forces

3.1 Introduction

3.1.1 Error Source and Classification of the Cable Networks

As shown in Fig. 3.1, a typical tension truss antenna comprises a supporting truss and a pair of mirror cable networks mutually tensioned by low-stiffness tie springs. The cable networks consist of a front net, a rear net, a metal mesh attached to the front net, and a tension tie spring (also known as a tension array) between the front and rear net [23]. When the antenna is fully deployed, the tension array will tighten the nodes corresponding to the front net and the rear net, thus maintaining the parabolic configuration and the necessary overall stiffness of the structure. The metal mesh is attached to the front net and the parabolic surface is fitted with triangular surfaces. The electromagnetic transmitter/receiver locates at the focus of the parabolic surface. Compared with rigid surfaces, the flexibility of cable networks leads to a great challenge in realizing parabolic configuration with high precision. Applications demand surface errors smaller than 1/50th of the wavelength [35, 36]. As a quantitative example, an S-band mission with a frequency of 4.0 GHz (wavelength 75 mm) requires a shape error as small as 1.5 mm. Therefore, it is of great significance to study the factors that affect the mesh precision and propose the design method of the cable networks with high precision mesh.

In addition to the systematic errors of the cable networks due to the geometric mesh division, the manufacturing errors of cables, bars and tie springs will also be introduced in practice, as well as environmental errors such as on-orbit temperature changes and the difference between on-ground and on-orbit. Table 3.1 lists six factors that affect the accuracy of the shape precision in practice. On the one hand, these factors can be divided into two categories according to their sources, namely,

© The Author(s), under exclusive license to Springer Nature Singapore Pte Ltd. 2026

K. Fu, *Design and Prediction of In-orbit Dynamic Behavior of Deployable Mesh Antennas*, Springer Theses, https://doi.org/10.1007/978-981-92-0140-2_3

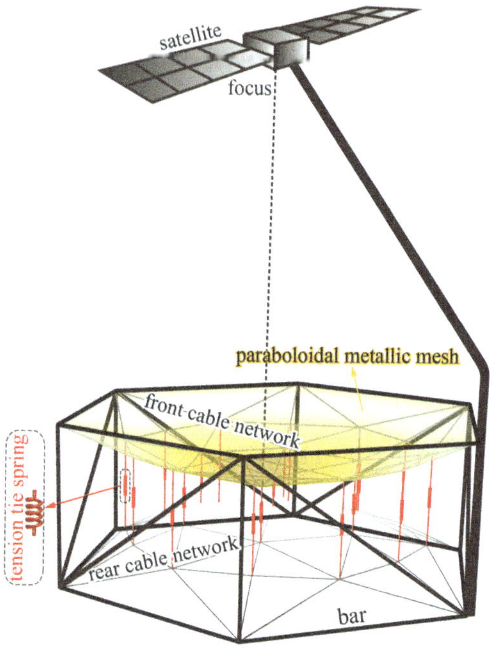

Fig. 3.1 Cable networks of a mesh antenna. Adapted with permission from [119]

Table 3.1 Error source and classification affecting the accuracy of profile

		Length error	Force error
System error	Geometric division error		
Manufacturing error	Cable length	√	
	Bar length	√	
	Spring ties		√
Environmental error	Gravity		√
	Temperature	√	√

manufacturing error and environmental error. On the other hand, the surface error can be divided into two types, namely, the length error and the force error.

3.1.2 Robust Cable Networks

As shown in Fig. 3.1, a typical tension truss antenna comprises a supporting truss and a pair of mirror cable networks mutually tensioned by low-stiffness tie springs. A significant advantage of the cables is that, if tensioned, they are mechanically equivalent to, but lighter than, bars. Usually, the supporting truss and cables are designed to form a statically determinate structure. Owing to this design, the shape of the cable network is determined solely by the lengths of the structure members with a given topology. Hence, if the length error is negligible, the cable nodes locate on the ideal surface.

The effect of length errors on the surface error has been studied extensively, and it has been suggested that shape precision can be improved by adjusting the lengths of a set of preselected cables. Nevertheless, the prerequisites of the previous studies might be violated; for example, cables might go slack if a notable variance of the tie spring forces occurs, which in turn would cause significant surface deformation. In practice, force variance is remarkable and has multiple sources, such as manufacturing errors, in-orbit working environment–orbital thermal and satellite spinning, and on-ground measurements of surface error.

The importance of ensuring that all cables are tensioned within the range of force uncertainties has long been recognized. It was first suggested that manufacturing errors be controlled and applications be limited to avoid large force variations. Later, Mitsugi and Tibert proposed the use of Monte Carlo simulations to check whether cables slackened due to deviations of the tie forces. However, the reverse problem of calculating the allowable deviation of the forces that can guarantee tautness of all of the cables has not yet been investigated, to the best of our knowledge, in the literature. This might be due to the lack of a precise definition of force tolerance and an efficient method to formulate and solve tolerance problems. As a result, no further effort has been made to optimize the cable network to obtain increased force tolerance.

The aim of this chapter is to design a robust cable network for statically determinate tension truss antennas that accounts for the variation in tie forces. First, we introduce a quantitative definition of the force tolerance of a truss antenna. Next, an analytical method of formulating and calculating force tolerance is developed. This approach is then exploited to optimize the cable network to achieve a larger force tolerance. By using 2 m front-fed and 6 m offset-fed mesh antennas as examples, we demonstrate that, after optimization, force tolerances are enlarged up to about four and two times, respectively, as compared to the original designs. Such improvements are beneficial in reducing manufacturing errors, expanding the acceptable working environment, and covering the gravity difference between on-ground testing and in-orbit usage. The proposed method is simple and efficient, and it can be applied to the design of mesh antennas.

3.2 Definition and Calculation of Force Tolerance

In engineering, a truss consists only of bars and cables. A statically determinate truss is one in which the internal forces and reactions at equilibrium are determined solely by the static equations when subjected to external loads. This study deals with the force tolerance problem of a statically determinate truss antenna that is composed of a supporting ring truss and cable networks. A truss cannot be statically determinate unless it meets Maxwell's rule [61], which states that the number of truss members and mechanical constraints equals the number of equilibrium equations. Figure 3.2 shows two two-dimensional trusses designed to locate nodes B and D on a parabolic curve. They each consist only of cables, meet Maxwell's rule, and are statically determinate, given that all cables are tensioned by external forces.

3.2.1 *Taut Region*

The requirement for external loads to tension all cables was studied by considering a general statically determinate truss with n external forces and c cables. Assuming that each external force is along a specific line but has variable values, positive directions can be defined; negative force values thus refer to the inverse direction of the forces. The values of external forces span an n-dimensional linear space.

In the static analysis presented hereafter, all cables are first considered as bars capable of bearing both tension and compression. As such, any force vector in the force space corresponds to a unique state of internal forces at equilibrium. All force value vectors that result in positive internal forces of all cables (that is, tension) enclose a so-called taut region.

The trusses in Fig. 3.2, which have two different topologies, are used as examples to illustrate the definition of the taut region. For the first one, for which node E coincides with the intersection point of lines AC and BD, once $f_1 > 0$ and $f_2 > 0$, all of the cables are tensioned, regardless of geometry. Hence, the taut region is the

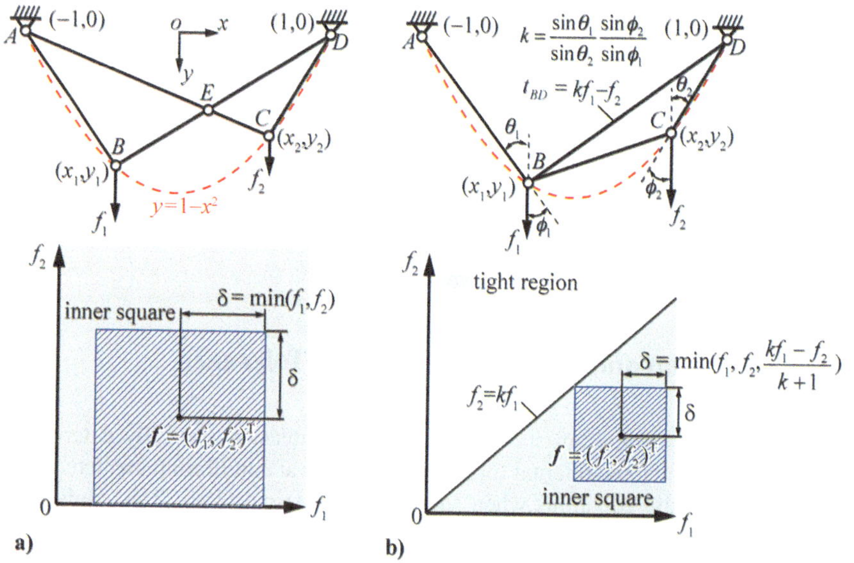

Fig. 3.2 Statically determinate truss, taut region, and absolute force tolerance. Adapted with permission from [119]: (**a**) the first truss; (**b**) the second truss

entire first quadrant of the f_1f_2 plane, as sketched in Fig. 3.2a. For the second truss, besides $f_1 > 0$ and $f_2 > 0$, another condition that

$$f_2 < kf_1, \; k = \frac{\sin\theta_1 \, \sin\phi_2}{\sin\theta_2 \, \sin\phi_1} \tag{3.1}$$

is demanded to tension cable *BD*, where k is a dimensionless factor determined by the truss geometry. Therefore, its taut region is a triangular region in the first quadrant, as shown in Fig. 3.2b. The shape of the taut region changes along with the truss geometry. When point *C* approaches *D* or point *B* approaches *A*, the region expands to cover the entire first quadrant resulting from $k \to \infty$.

3.2.2 Definition of Force Tolerance

3.2.2.1 Absolute Force Tolerance

For a general statically determinate truss, assuming that a force vector $\boldsymbol{f}$ is in the taut region, the taut condition of the i-th cable ($i = 1, 2, \cdots, c$) is first independently solved, and then the taut condition of the entire truss is determined. For the i-th cable, deviate $\boldsymbol{f}$ to $\boldsymbol{f} + \boldsymbol{\Delta}$, where each component Δ_α, $\alpha = 1, 2, \cdots, n$ of $\boldsymbol{\Delta}$ can vary freely from $-\delta_i^f$ to δ_i^f. Then, δ_i^f denotes the absolute force tolerance of the i-th cable if the internal force of the i-th cable $\delta_i^{f+\Delta}$ can reach zero but is never negative; that is,

$$\begin{aligned} & t_i^{\mathbf{f}+\Delta} \geq 0 \\ & -\delta_i^{\mathbf{f}} \leq \Delta_\alpha \leq \delta_i^{\mathbf{f}}, \alpha = 1,2,\cdots,n \end{aligned} \tag{3.2}$$

The minimum of all δ_i^f, or $\delta^f = \min\left(\delta_1^f, \delta_2^f, \cdots, \delta_c^f\right)$, is defined as the absolute force tolerance of the entire truss with a given $\boldsymbol{f}$. Based on the preceding definition of the absolute force tolerance, δ^f can be geometrically understood as half the length of the inner cube of the n-dimensional taut region with center point $\boldsymbol{f}$. It is a function of the applied force vector $\boldsymbol{f}$, the topology, and the geometry of the truss.

Figure 3.2 shows the absolute force tolerances of two different trusses. The applied force vector is $\boldsymbol{f} = (f_1, f_2)^{\mathrm{T}}$. For the first truss, tensioning each cable requires $f_1 > 0$, $f_2 > 0$. Therefore, the absolute force tolerance of each cable is $\min(f_1, f_2)$, regardless of the geometry. Then, the absolute force tolerance of the entire truss δ^f is $\min(f_1, f_2)$. For the second truss, the tolerance of cables *AB*, *BC*, and *CD* is $\min(f_1, f_2)$; and the tolerance of cable *BD* is $(kf_1 - f_2)/(k + 1)$, so $\delta^f = \min(f_1, f_2, (kf_1 - f_2)/(k + 1))$, depending on the geometry because k is a function of geometry.

3.2.2.2 Absolute Force Tolerance

If the force $\boldsymbol{f}$ is scaled by a constant, then δ^f is also scaled. To eliminate the effect of force magnitude, the relative force tolerance δ^{rel} is introduced, which is defined as the ratio of the absolute force tolerance δ^f to the force magnitude. Here, the magnitude is measured in terms of the infinity norm of $\boldsymbol{f}$ or

$$\delta^{\text{rel}} = \frac{\delta^f}{f_\infty} = \frac{\min\left(\delta_1^f, \delta_2^f, \cdots, \delta_c^f\right)}{\max\limits_{i=1,2,\cdots,n} |f_i|} \tag{3.3}$$

Because $\boldsymbol{f}$ can tension all cables, this definition implies that the possible relative force tolerance should be larger than zero and no larger than one, or $\delta^{\text{rel}} \in [0,\ 100\%]$.

3.2.3 Formulation of the Force Tolerance of a Tension Truss Antenna

For a more complicated three-dimensional tension truss antenna, the force tolerances are formulated by using the antenna shown in Fig. 3.1 as an example. In a static analysis, the cable networks can be modeled with cable elements, and the supporting truss members can be simplified as bars because there is no load applied between any two nodes of the truss members. Due to the mirror symmetry of the front and back networks, we consider a truss to be composed of cables of the front network and bars of the top circular ring. In Fig. 3.3a, the nodes that link to any bars are marked as solid circles and are called bar nodes. The nodes that only link to cables are marked as empty circles and are called cable nodes. The local coordinate system $oxyz$ is chosen such that the top ring lies on the xy plane and the center of the ring locates at the local origin o. Each cable node is subjected to a tie force along the z axis. The positive direction of the forces is along the negative z direction. Each bar node is confined to the xy plane. Meanwhile, a bar node is fixed in both the x and y directions, and its opposite node is restricted in the y direction. Then, the bars and cables together form a statically determinate truss, implying that Maxwell's rule is met, i.e.,

$$0 = 3(n+b) - (b+c+b+3) = 3n + b - c - 3 \tag{3.4}$$

where b is the number of bars as well as the number of bar nodes, n is the number of cable nodes as well as the number of tie forces, and c is the number of cables.

For a given circular ring or number of bars b, there is an infinite number of cable networks with combinations of n cable nodes and c cables that could meet Maxwell's rule as shown in Eq. (3.4). Several net examples with different b, n, and c are given in Ref. [63]. For the case of $b = 6$, Fig. 3.3 shows two nets: the first

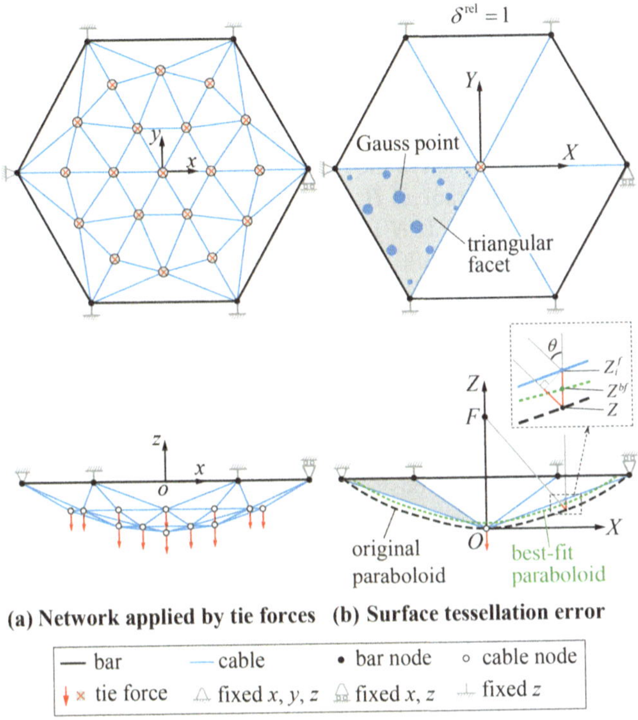

Fig. 3.3 Three-dimensional truss with a top ring and a front network. Adapted with permission from [119]: (**a**) network applied by tie forces; (**b**) surface tessellation error

one has $n = 19$ cable nodes and $c = 60$ cables, and the second has only $n = 1$ cable nodes and $c = 6$ cables. They both satisfy Eq. (3.4), and both are statically determinate.

There are two steps for calculating the force tolerance of a tension truss antenna. The first is computing the internal cable tensions when the truss is enforced by a given tie force vector $\boldsymbol{f}$; this is a simple but essential step. The second is calculating the force tolerance of each cable.

3.2.3.1 Cable Tension of a Truss

For the currently studied truss, the static equations can be established by analyzing the equilibrium conditions of each node. There are $n + b$ nodes, so there are $N = 3(n + b)$ equations. They can be written in matrix-vector form

$$\boldsymbol{H}_{N\times N}\boldsymbol{s}^{f}_{N\times 1} + \boldsymbol{D}_{N\times n}\boldsymbol{f}_{n\times 1} = \boldsymbol{0} \tag{3.5}$$

where the subscripts represent the dimension of the vectors or matrices; $\boldsymbol{f} = [f_1, f_2, \cdots, f_n]^{\mathrm{T}}$ is a vector for which the components are the values of the tie force; $\boldsymbol{D}$ is a constant direction matrix converting $\boldsymbol{f}$ to the three-dimensional external forces on all nodes; $\boldsymbol{s}^f$ is referred to as the state vector and is composed of the internal forces and reactions; and $\boldsymbol{H}$ is a matrix assembled from equilibrium equations, which is completely determined by the topology, geometry, and constraints of the truss. Checking the rank of $\boldsymbol{H}$ allows determination of whether the truss is statically determinate [120]. If $\boldsymbol{H}$ is a full rank or invertible matrix, the truss is statically determinate, and each external force $\boldsymbol{f}$ corresponds to a unique state vector $\boldsymbol{s}^f$.

For convenience, the c internal forces of all cables in the state vector $\boldsymbol{s}^f$ are denoted as $\boldsymbol{t}^f$. Given $\boldsymbol{H}$, both $\boldsymbol{s}^f$ and $\boldsymbol{t}^f$ are linear functions of $\boldsymbol{f}$. Consider a unit-force vector $\boldsymbol{f}^\alpha$, $\alpha = 1, 2, \cdots, n$, for which the α-th component is one and all other components are zero. Then, there is a corresponding state vector $\boldsymbol{s}^\alpha = \boldsymbol{H}^{-1}\boldsymbol{D}\boldsymbol{f}^\alpha$ and a cable force vector $\boldsymbol{t}^\alpha$. Defining $\boldsymbol{T} = [t^1, t^2, \cdots, t^n]$, then the cable force vector $\boldsymbol{t}^f$ is a linear sum of $\boldsymbol{t}^\alpha$, $\alpha = 1, 2, \cdots, n$, which can be expressed as

$$\boldsymbol{t}^f = \boldsymbol{T}\boldsymbol{f} = f_1\boldsymbol{t}^1 + f_2\boldsymbol{t}^2 + \cdots + f_n\boldsymbol{t}^n \tag{3.6}$$

In general, some components of $\boldsymbol{t}^\alpha$ are positive and some are negative or zero. Therefore, a general $\boldsymbol{f}$ may not tension all cables.

3.2.3.2 Force Tolerances of a Truss

Consider a force vector $\boldsymbol{f}$ in the taut region. If $\boldsymbol{f}$ deviates to $\boldsymbol{f} + \boldsymbol{\Delta}$, with $\boldsymbol{\Delta} = [\Delta_1, \Delta_2, \cdots, \Delta_n]^{\mathrm{T}}$, then the internal force of the i-th cable deviates linearly with respect to $\boldsymbol{\Delta}$ as

$$t_i^{f+\Delta} = t_i^f + \sum_{\alpha=1}^{n} t_i^\alpha \Delta_\alpha \tag{3.7}$$

For the i-th cable, assume that $\Delta_\alpha \in \left[-\delta_i^f, \delta_i^f\right]$, $\alpha = 1, 2, \cdots, n$, the minimum value of $t_i^{f+\Delta}$ will then be reached at

$$\Delta_\alpha = -\delta_i^f \times sign\left(t_i^\alpha\right) \tag{3.8}$$

in which sign(x) is the sign function defined in Eq. (2.11). This means that the tie forces that can result in negative tension on the i-th cable uniformly deviate to the maximum positive value δ_i^f, whereas the tie forces that can result in positive tension on the i-th cable uniformly deviate to the maximum negative value $-\delta_i^f$. According to the definition of the absolute force tolerance of the i-th cable, the minimum value of $t_i^{f+\Delta}$ is zero. Substituting Eq. (3.8) into Eq. (3.7) results in

$$\delta_i^f = \frac{t_i^f}{\sum_{\alpha=1}^{n} \mathrm{abs}\left(t_i^\alpha\right)} \tag{3.9}$$

where abs(x) is the absolute value of x. Consequently, the absolute force tolerance of the entire truss δ^f is the minimum of all δ_i^f, expressed as

$$\delta^f = \min\left(\delta_1^f, \delta_2^f, \cdots, \delta_c^f\right) \tag{3.10}$$

Then, the relative force tolerance can be expressed as

$$\delta^{\mathrm{rel}} = \frac{\delta^f}{f_\infty} = \frac{\min\left(\delta_1^f, \delta_2^f, \cdots, \delta_c^f\right)}{f_\infty} \tag{3.11}$$

3.2.4 *Simple Analytical Example*

Taking the cable network shown in Fig. 3.3b as an example, we calculate its relative force tolerance. There is only one cable node ($n = 1$) at the center and one external force $\boldsymbol{f}$. In statics, all cable tensions are positive while $f > 0$ and will therefore, according to Eqs. (3.9) and (3.11), $\delta^{\mathrm{rel}} = 100\%$; it reaches the maximal boundary of its range. This result says that the relative force tolerance will reach its global maximum of one if the net tessellates the paraboloidal surface with only one cable node and $b = c$ triangular facets.

3.3 Optimization of the Relative Force Tolerance

Hereafter, to obtain a robust design of the tension truss antennas with respect to the applied forces, the relative force tolerance of a given truss will be improved by optimizing the distribution of the applied tie forces and/or the geometry of the truss. For simplicity, the infinite norm of $\boldsymbol{f}$ is set to be one, or $\|\boldsymbol{f}\|_\infty = 1$, so that $\delta^{\mathrm{rel}} = \delta^f$.

3.3.1 *Optimization of the Tie Forces*

If the topology and geometry of a truss are given and not changeable, then maximizing the relative force tolerance of a truss (that is, maximizing the minimum value of the force tolerances of all cables) can be achieved only by varying the distribution of the tie forces. This is a typical maximum–minimum optimization problem, which may be formulated using a bound formulation [121, 122] as follows

$$\begin{array}{ll} \text{Find} & \beta, \boldsymbol{f} \\ \textit{maximize} & \beta \\ \text{subject to} & \beta - \delta_i^f(\boldsymbol{f}) \le 0,\ i = 1, 2, \cdots, c \\ & \boldsymbol{t}^f(\boldsymbol{f}) > \mathbf{0} \\ & 0 \le f_\alpha \le 1,\ \alpha = 1, 2, \cdots, n \end{array} \tag{3.12}$$

where the constraint $\boldsymbol{t}^f > \mathbf{0}$ means that each component of $\boldsymbol{t}^f$ is positive. The design variables include a dummy variable β and tie forces $\boldsymbol{f}$, where β is introduced to ensure that the objective function of the maximum–minimum problem is smooth and continuous. Maximizing β is equivalent to maximizing the relative force tolerance. Both the objective function and the constraints are linear functions of the design variables because δ_i^f is a linear function of $\boldsymbol{f}$. Therefore, Eq. (3.12) is a linear programming problem [123], and its global optimum solution $\beta_{\text{opt}}^{\text{linear}}$ and $\boldsymbol{f}_{\text{opt}}^{\text{linear}}$ can be obtained by a standard linear programming method, e.g., the simplex method.

3.3.2 *Simultaneous Optimization of the Tie Forces and the Geometry of the Truss*

If the topology of a truss is given and fixed while the geometry of the truss can be changed, then the relative force tolerance of a truss can be optimized by simultaneously varying the distribution of the tie forces and the geometry of the truss. The geometry could be described by a position vector $\boldsymbol{R} = [x_1, y_1, x_2, y_2, \cdots, x_n, y_n]^{\mathrm{T}}$ consisting of the coordinates x and y of all cable nodes, assuming that the nodes are located on a paraboloid surface $z = z(x, y)$. Then, the maximum–minimum problem of optimizing the relative force tolerance could be expressed using the bound formulation as follows

$$\begin{array}{ll} \text{Find} & \beta, \boldsymbol{f}, \boldsymbol{R} \\ \textit{maximize} & \beta \\ \text{subject to} & \beta - \delta_i^f(\boldsymbol{f}, \boldsymbol{R}) \le 0,\ i = 1, 2, \cdots, c \\ & \boldsymbol{t}^f(\boldsymbol{f}, \boldsymbol{R}) > \mathbf{0} \\ & 0 \le f_\alpha \le 1,\ \alpha = 1, 2, \cdots, n \end{array} \tag{3.13}$$

The design variables include a dummy variable β, tie forces $\boldsymbol{f}$, and geometry positions $\boldsymbol{R}$.

3.3.3 Constraint of Tessellation Error

In the preceding optimization, any adjustment to the geometry position $\boldsymbol{R}$ will change the tessellated surface error, which is a concern in antenna design. Therefore, it is better to introduce additional constraints to limit the change. The simplest way is to limit the length of the i-th cable within an interval $[l_{i,\min}, l_{i,\max}]$, i.e.,

$$l_{i,\min}^2 \le \left(x_p - x_q\right)^2 + \left(y_p - y_q\right)^2 + \left(z_p - z_q\right)^2 \le l_{i,\max}^2, i = 1,2,\cdots,c \tag{3.14}$$

where the i-th cable has two nodes: the p-th and q-th nodes. However, a more intuitive method is to directly limit the tessellation error. Herein, the half-path-length best-fit RMS error $e_{\text{hp}}^{\text{bf}}$ is adopted to evaluate the tessellation error because it relates directly to the actual antenna performance. This involves two concepts, the half-path-length RMS error and the best-fit paraboloid.

As shown in Fig. 3.3b, the half-path-length RMS error [77] is defined as

$$e_{\text{hp}} = \sqrt{\sum_{i=1}^{N_f} \int_{A_i} \left(\frac{e_i}{2}\right)^2 \text{d}A \Big/ \sum_{i=1}^{N_f} \int_{A_i} \text{d}A} \tag{3.15}$$

in which

$$e_i = \left(1 + \cos\theta\right)\left(Z_i^f - Z\right) \tag{3.16}$$

where $Z = (X^2 + Y^2)/(4F)$ is the target paraboloid surface with focal length F. N_f is the number of triangular facets. For the i-th triangular facet, A_i is the projected region of the facet on the XY plane; e_i is the path length error between the facet and the paraboloid surface, where Z_i^f is the Z coordinate of the triangular facet; and θ is the angle between the boresight direction and the direction from a point on the reflector to the focus. The integration over each facet can be calculated via Gauss–Legendre quadrature for triangles [124, 125]; and the distribution of Gauss points with the corresponding weights, using 16 points as an example, is illustrated in Fig. 3.3b.

The best-fit paraboloid [126] is the one that minimizes the tessellation error by shifting the original paraboloid along the Z direction by ΔZ and adjusting the focal length to F^{bf} [20, 32], which can be expressed as

$$Z^{\text{bf}} = Z^{\text{bf}}\left(X,Y\right) = \left(X^2 + Y^2\right)/\left(4F^{\text{bf}}\right) + \Delta Z \tag{3.17}$$

The half-path-length best-fit RMS error $e_{\text{hp}}^{\text{bf}}$ is similar to Eq. (3.15) except Z is replaced by the best-fit paraboloid Z^{bf}, expressed as

$$e_{\mathrm{hp}}^{\mathrm{bf}} = \sqrt{\sum_{i=1}^{N_f} \int_{A_i} \left(\frac{e_i^{\mathrm{bf}}}{2} \right)^2 \mathrm{d}A \Big/ \sum_{i=1}^{N_f} \int_{A_i} \mathrm{d}A} \tag{3.18}$$

in which

$$e_i^{\mathrm{bf}} = (1+\cos\theta)\left(Z_i^f - Z^{\mathrm{bf}}\right) \tag{3.19}$$

The parameters F^{bf} and ΔZ of the best-fit paraboloid can be analytically obtained if θ^{bf} is assumed to be the same as θ of the original paraboloid, or it can be calculated by using standard optimization algorithms. In the current study, ΔZ and $\Delta F = F^{\mathrm{bf}} - F$ are small, so we assume $\theta^{\mathrm{bf}} = \theta$ and the best-fit paraboloid can be calculated analytically. During optimization, a constraint can be optionally added to limit $e_{\mathrm{hp}}^{\mathrm{bf}}$ no larger than a certain threshold value $e_{\mathrm{threshold}}$, expressed as

$$e_{\mathrm{hp}}^{\mathrm{bf}} \le e_{\mathrm{threshold}} \tag{3.20}$$

where $e_{\mathrm{hp}}^{\mathrm{bf}}$ can be calculated through Eq. (3.18).

3.3.4 *Optimization Procedure of Force Tolerance*

The optimization problem formulated by Eqs. (3.12), (3.13), and (3.20) is nonlinear with respect to position vector $\boldsymbol{R}$. In the next section, we show that this nonlinear problem is not guaranteed to be convex. Therefore, it is difficult to achieve the global optimal solution. The problem is solved here by using the sequential quadratic programming (SQP) algorithm [127]. The algorithm requires a group of initial values $\beta_0^{\mathrm{non}}, f_0^{\mathrm{non}}, \boldsymbol{R}_0^{\mathrm{non}}$ for iteration. In our study, $\boldsymbol{R}_0^{\mathrm{non}}$ is set as the initial position vector of the cable nodes $\boldsymbol{R}_0$, β_0^{non} and f_0^{non} are set as the optimum solution $\beta_{\mathrm{opt}}^{\mathrm{linear}}$ and $f_{\mathrm{opt}}^{\mathrm{linear}}$ of the linear programming described by Eq. (3.12) with given $\boldsymbol{R}_0$. The procedure of enlarging the relative force tolerance through optimization is illustrated in Fig. 3.4.

3.4 Examples of Robust Design

Two tension truss antennas are considered in this section to actualize the proposed robust design against variation in the tie forces. One is a proof of concept using a simple truss, and the other is more complex and is meant for practical applications.

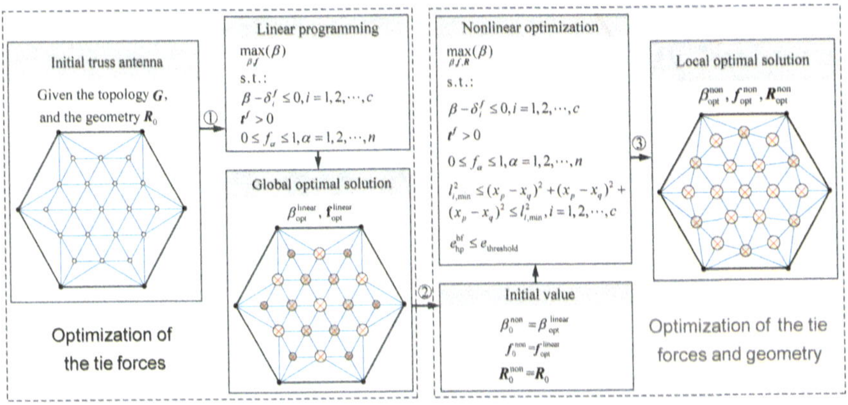

Fig. 3.4 Optimization procedure of enlarging the relative force tolerance. Adapted with permission from [119]

3.4.1 *Robust Design of a Six-Bay Two-Meter Front-Fed Truss Antenna*

3.4.1.1 Optimization of the Relative Force Tolerance

Figure 3.5a shows a frontfed truss antenna; its paraboloid surface is axisymmetric with a focal length of $F = 7.5$ m and an aperture diameter of $D = 2$ m. The top of the antenna contains 6 bars, 19 cable nodes, and 60 cables. According to Eqs. (3.4) and (3.5), the structure not only meets Maxwell's rule but also has a full rank $\boldsymbol{H}$. Therefore, the truss is statically determinate.

Although there are 19 cable nodes, the positions of these cable nodes can be fully determined by three independent parameters $\boldsymbol{r} = [r_1, r_2, r_3]^{\mathrm{T}}$ due to their geometric symmetry. However, in the following optimization, the tie forces $\boldsymbol{f}$ are not assumed to be rotationally symmetrical because each component can vary independently.

For a geometry with $\boldsymbol{r} = [0.333,\ 0.667,\ 0.667]^{\mathrm{T}}$ m, if the tie forces are uniformly set to be one (i.e., $\boldsymbol{f} = [1, 1, \cdots, 1]^{\mathrm{T}}$ N), then the relative force tolerance $\delta^{\mathrm{rel}} = 7\%$ can be obtained. After the linear programming optimization of Eq. (3.12), the tolerance is enlarged to $\delta^{\mathrm{rel}} = 11\%$ by varying only $\boldsymbol{f}$ while keeping $\boldsymbol{r}$ unchanged, as shown in Fig. 3.5b. Then, the network geometry (i.e., $\boldsymbol{r}$) is also optimized to further improve the tolerance. During optimization, the minimum length of all cables is limited to $l_{\min} = 0.1$ m, whereas no constraint is added on the half-path-length best-fit RMS error. By using the nonlinear optimization described by Eqs. (3.13) and (3.14), a converged solution can be achieved, as shown in Fig. 3.5c, where the relative force tolerance is expanded to $\delta^{\mathrm{rel}} = 22\%$, the optimum geometry is $\boldsymbol{r}_{\mathrm{opt}}^{\mathrm{non}} = [0.491, 0.900,\ 0.697]^{\mathrm{T}}$ m, and the optimum tie forces $\boldsymbol{f}_{\mathrm{opt}}^{\mathrm{non}}$ range from 0.63 to 1 N, as shown in Fig. 3.5c.

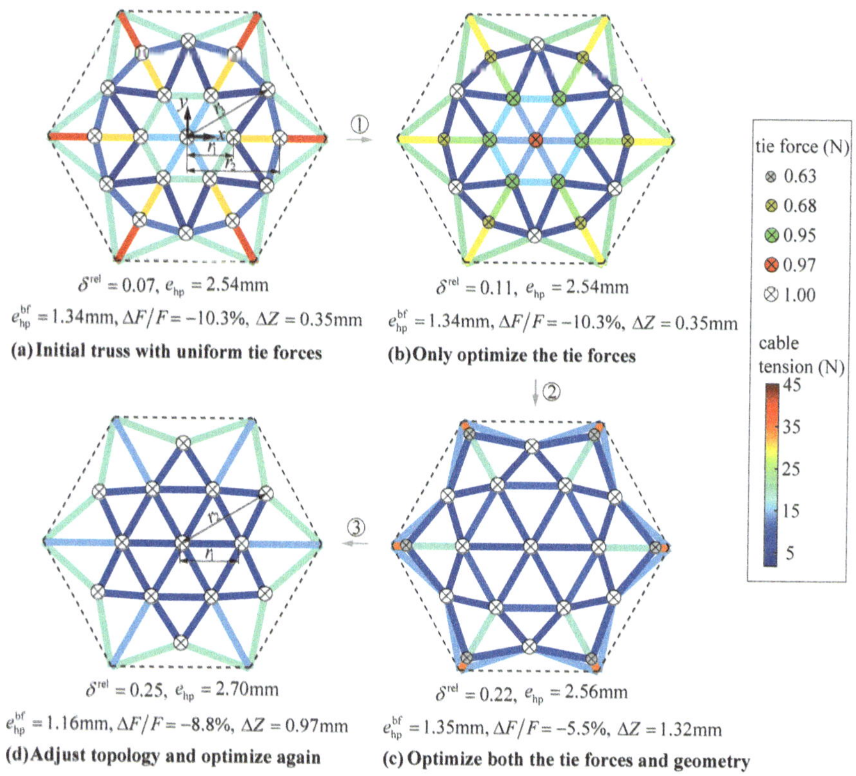

Fig. 3.5 Optimization of a six-bay 2 m frontfed truss. (**a**) initial truss with uniform tie forces; (**b**) only optimize the tie forces; (**c**) optimize both the tie forces and geometry; (**d**) adjust topology and optimize again. Adapted with permission from [119]

More interestingly, the optimum result illustrated in Fig. 3.5c shows that the six outermost cable nodes tend to coincide with the bar nodes. We thus infer that the employed topology is not optimal; a better truss should have the six outermost cable nodes coincide with the bar nodes, as shown in Fig. 3.5d. The new truss will remain statically determinate, and the geometry of the cable network can be determined by two independent parameters: $\boldsymbol{r} = [r_1, r_2]^{\mathrm{T}}$. By optimizing both $\boldsymbol{r}$ and $\boldsymbol{f}$, a solution converges to $\boldsymbol{r}_{\mathrm{opt}}^{\mathrm{topo}} = [0.432, 0.696]^{\mathrm{T}}$ m and $\boldsymbol{f}_{\mathrm{opt}}^{\mathrm{topo}} = [1, 1, \cdots, 1]^{\mathrm{T}}$ N, as shown in Fig. 3.5d. The final obtained tolerance is $\delta^{\mathrm{rel}} = 25\%$, which is about four times the original value.

The aforementioned force tolerance analysis procedure is validated through a stress analysis conducted by using commercial finite element software: ABAQUS 6.14. The two-dimensional bar element (Type: T3D2) was adopted which can withstand both tension and pressure. The stiffness of the elements EA is set to be 2×10^5 N to ensure that the elastic deformation of the cables can be ignored.

According to the final optimization results, all the cables are tensioned under the optimum tie forces $\boldsymbol{f}_{\mathrm{opt}}^{\mathrm{topo}} = [1, 1, \cdots, 1]^{\mathrm{T}}$ N. Taking the $\boldsymbol{f}_{\mathrm{opt}}^{\mathrm{topo}}$ as a reference, a group of force variation are applied and the tie forces become $\boldsymbol{f} = \boldsymbol{f}_{\text{opt}}^{\text{topo}} + \Delta$. The allowable change of each force component of the tie forces is $-\delta^{\text{rel}} \le \Delta_\alpha \le \delta^{\text{rel}}$, $\alpha = 1, 2, \cdots, 13$, where the relative force tolerance $\delta^{\text{rel}} = 0.25$ has been given by the optimization results. The optimization results also show that the relative force tolerance of the 15th cable (the cable number is shown in Fig. 3.6b) is one of the minimum relative force tolerances of all the cables. The critical conditions when the tension of the 15th cable is 0 is as follows

$$\Delta_\alpha = \left(\delta^{\text{rel}} + \varepsilon\right) \times \begin{cases} 1, \alpha = 2,6,7,12,13 \\ -1, \alpha = 1,3,4,5,8,9,10,11 \end{cases}, \ \varepsilon = 0 \tag{3.21}$$

The results of ABAQUS simulation verify the above conclusions. As shown in Fig. 3.6a, when $\varepsilon = -0.007 < 0$, the force deviation of the tie forces is controlled within the relative force tolerance δ^{rel}, and ABAQUS results show that all the cables are taut. As shown in Fig. 3.6b, when $\varepsilon = 0.013 > 0$, the force deviation of the tie forces exceeds the range of relative force tolerance δ^{rel}. At this time, ABAQUS results show that the internal forces of the 15th and 23rd cables are −0.06 N, indicating that these two cables are slack. Therefore, the results of ABAQUS validation show that when the tie force error is controlled within the relative force tolerance, the tension of all cables can be guaranteed. When the error of the tie forces exceeds

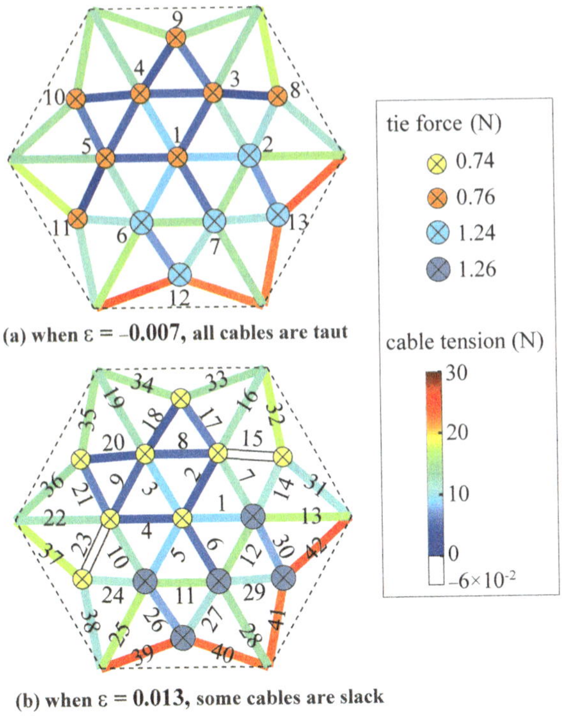

(a) when ε = −0.007, all cables are taut

(b) when ε = 0.013, some cables are slack

Fig. 3.6 Validation of optimization results: (**a**) when ε = −0.007, all cables are taut; (**b**) when ε = 0.013, some cables are slack

the relative force tolerance, some of the cables are slack. This is consistent with the expectation, which indicates that the analytical solution method of force tolerance and the optimization algorithm of increasing force tolerance are correct.

3.4.1.2 Discussions on the Optimized Results

Comparing the final optimized topology in Fig. 3.5d with the initial topology in Fig. 3.5a, the relative force tolerance increased, whereas the number of cable elements decreased. This naturally raised two questions: first, whether the tessellation error was remarkably increased in the final optimized configuration because fewer elements were used; and second, whether optimization tended to reduce the rings of the networks to increase relative force tolerance.

The first question is easy to answer by numerically checking the tessellation errors through Eqs. (3.15) and (3.18). For the integration, a Gauss–Legendre quadrature over a triangle with orders of 4 and 16 integral points is adopted to ensure numerical integral accuracy. The results show that the half-path-length RMS error $e_{\mathrm{hp}} = 2.54$ mm in the initial configuration and increases by 6.3% to 2.70 mm in the final optimized configuration. This increment is expected because the optimized configuration has a smaller number of tessellation facets. In addition, the half-path-length best-fit RMS errors $e_{\mathrm{hp}}^{\mathrm{bf}}$ are also checked. It reduces a little bit from 1.34 mm in Fig. 3.5a to 1.16 mm in Fig. 3.5d. This result is counterintuitive because Fig. 3.5d has larger triangles than Fig. 3.5a. The reason might be that the number of cable nodes is few, resulting in the ΔF and ΔZ values of the best-fit paraboloids being quite different in these two configurations.

The second question is difficult to answer but worth thinking about. In our opinion, the answer is yes; the trend is consistent with the results shown in Fig. 3.3b, in which the relative force tolerance reaches the global maximum of one with a minimal number of cable nodes of one and a minimal number of triangular facets of b. However, optimization does not always tend to delete the rings of a network and may not approach the global optimum for several reasons; for instance, cable lengths are constrained by Eq. (3.14) and optimization might be nonconvex so that the algorithm converges to a local optimum rather than the global optimum, as proven in the following context.

3.4.1.3 Nonconvex Nature of the Nonlinear Optimization Formulation

By using the truss topology shown in Fig. 3.5d as an example, it is shown that the nonlinear optimization formulated by Eqs. (3.13) and (3.14) might be nonconvex. In this example, the design variables include a dummy variable β, tie forces $\boldsymbol{f}$, and geometry parameters $\boldsymbol{r}$. It is known that the optimum tie forces are uniform, i.e., $\boldsymbol{f}_{\mathrm{opt}}^{\mathrm{topo}} = [1,1,\cdots,1]^{\mathrm{T}}$ N. If $\boldsymbol{f} = \boldsymbol{f}_{\mathrm{opt}}^{\mathrm{topo}}$ is fixed, then only three design variables remain:

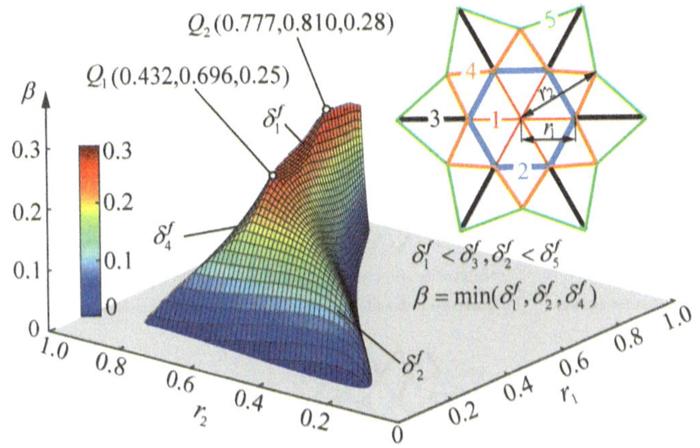

Fig. 3.7 Nonconvex nature of the nonlinear optimization. Adapted with permission from [119]

that is, β and $\boldsymbol{r} = [r_1, r_2]^{\mathrm{T}}$. β is the minimum value of $\delta_i^f(r_1, r_2)$, $i = 1, 2, \cdots, 42$, which can be intuitively plotted by surfaces in three-dimensional space.

Based on the axial symmetry of the shape, the force tolerance of each cable can be classified into five groups: $\delta_1^f, \delta_2^f, \delta_3^f, \delta_4^f$ and δ_5^f, as shown in Fig. 3.7a. A further analysis shows that $\delta_1^f < \delta_3^f, \delta_2^f < \delta_5^f$, so that $\beta = \min\left(\delta_1^f, \delta_2^f, \delta_4^f\right)$. The surfaces of $\delta_1^f, \delta_2^f, \delta_4^f$ are functions of (r_1, r_2) and are plotted in Fig. 3.7b. The intersected area is clearly not convex. There are two local optimum solutions: one is $Q_1(r_1, r_2, \beta) = (0.432, 0.696, 0.25)$, and the other is $Q_2(r_1, r_2, \beta) = (0.777, 0.810, 0.28)$. The previously obtained solution converges to Q_1. In fact, if the initial geometry of the truss $\boldsymbol{r} = [r_1, r_2]^{\mathrm{T}} = [0.6, 0.7]^{\mathrm{T}}$ for interaction, the optimum solution will converge to Q_2.

3.4.2 *Robust Design of a Thirty-Bay Offset-Fed Truss Antenna*

In real applications, the antennas are offset fed [128]. In this case, as shown in Fig. 3.8a, the reflector surface is a segment of a paraboloid intersected by a cylinder, for which the axis is offset from the paraboloid axis by $d = 4$ m. The surface is asymmetric, and the top ring is an ellipse with a semiminor axis of $r = 3$ m and a semimajor axis of $D/2 = 3.905$ m. It is not circular, as front-fed antennas are. Here, the top ring is meshed with 30 bars and the front net is tessellated with 61 cable nodes and 210 cables. The rear net is a mirror of the front net. This structure possesses a realistic application size and geometry. It is statically determinate and is applied with robust design.

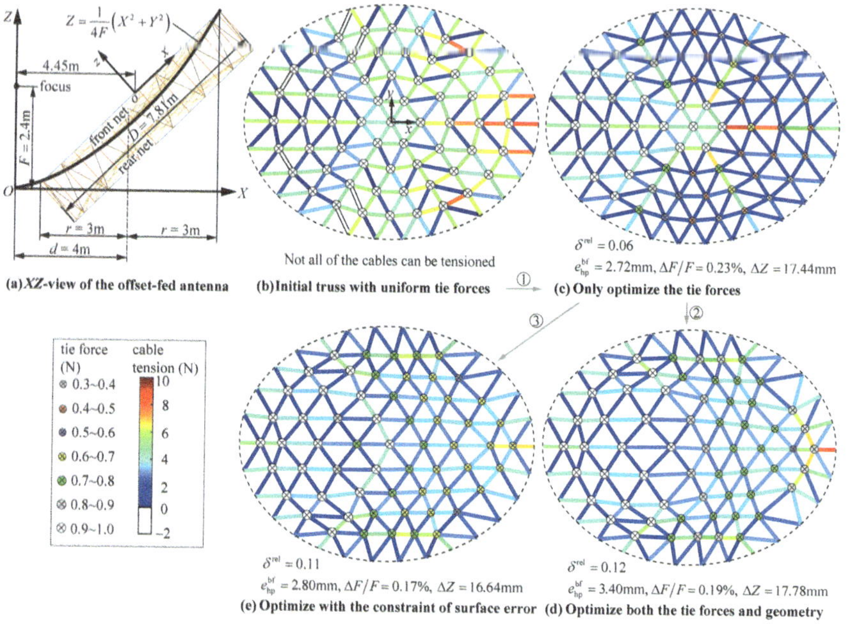

Fig. 3.8 Optimization of a 30-bay offset-fed truss. (**a**) *XZ*-view of the offset-fed antenna; (**b**) initial truss with uniform tie forces; (**c**) only optimize the tie forces; (**d**) optimize both the tie forces and geometry; (**e**) optimize with the constraint of surface error. Adapted with permission from [119]

A numerical calculation shows that not all cables in the initial network can be tensioned by a uniform tie force vector $\boldsymbol{f} = [1, 1, \cdots, 1]^{\mathrm{T}}$ N, as shown in Fig. 3.8b. A taut cable network is achieved only by optimizing the tie forces as shown in Fig. 3.8c, and the corresponding relative force tolerance $\delta^{\mathrm{rel}} = 6\%$. By optimizing both the tie forces and the geometry of the truss, during which the minimal cable length is set to be 0.6 m with no constraint on the best-fit tessellation error, the relative force tolerance further doubles to $\delta^{\mathrm{rel}} = 12\%$. The converged geometry is shown in Fig. 3.8d. It is asymmetric and irregular due to the offset of the feed. Other studies have revealed that an irregular network pattern could improve antenna performance [129]. However, the current improvement in the relative force tolerance sacrifices the half-path-length best-fit RMS error. The results show that $e_{\mathrm{hp}}^{\mathrm{bf}}$ increases from 2.72 to 3.40 mm, up to a total increment of 25%. This result is intuitive because the configuration in Fig. 3.8d has visibly larger triangles, and a consequently larger best-fit RMS error, based on the fact that the ΔF and ΔZ values of the best-fit paraboloids of the two configurations have no notable differences.

To keep the RMS error low, another optimization was conducted with an additional constraint of $e_{\mathrm{hp}}^{\mathrm{bf}}$ ″ 2.80 mm. The optimization converged to another geometry, shown in Fig. 3.8e. That configuration had $e_{\mathrm{hp}}^{\mathrm{bf}} = 11\%$ and $e_{\mathrm{hp}}^{\mathrm{bf}} = 2.80\,\mathrm{mm}$.

Compared with Fig. 3.8c and d, it possessed the advantages of both a large force tolerance and a relatively low RMS error.

Improving the force tolerance is beneficial to the on-ground measurement. For this 6 m antenna, the average tie force is about 3 N, and the mass of the front cable network with the attached metal mesh is around 2 kg. Assuming the density is uniform, the average gravity force on each cable node is $mg/\alpha = 2 \times 9.81/61 = 0.32$ N. If the relative force tolerance is 0.11, then the absolute force tolerance is $3 \times 11\ \% = 0.33$ N, which is just a bit larger than 0.32. This will guarantee the cables are tensioned on the ground without a suspension system, just as for the in-orbit case.

Moreover, the optimization is efficient. For the geometric optimization of Fig. 3.8e, a converged solution takes 665 s of CPU time with 42 iterations, conducted on an eight-core personal computer (Intel Core i7–4790 with 4.00 GHz). The adopted algorithm is SQP implemented in MATLAB 2015a, with the tolerance of the optimized variables of TolX = 1×10^{-6} and the termination tolerance of the objective function of TolFun = 1×10^{-3}.

3.5 Summary

For a tension truss antenna, previous work highlighted that variations in the tie forces, resulting from both manufacturing errors and in-orbit working environments, may cause cable slack; preventing cable slack is important for avoiding large surface deformations. However, no research has been done on either obtaining the allowable deviation of the forces that will guarantee tautness of all of the cables or designing a cable network that is robust against variance of the tie forces.

In this chapter, a novel methodology is presented for achieving such a robust network for statically determinate tension truss antennas. The methodology has two main parts: the first is an analytical formulation for calculating the force tolerances of a tension truss antenna, and the second is a formulation to optimize the force tolerance by varying the tie forces and the geometry of the truss. Applying this method to a 2 m front-fed mesh antenna reveals some insights on the nature of the formulation; for instance, the nonlinear optimization may not be convex, which may result in obtaining a solution that may be a local optimum rather than the global optimum.

Finally, the proposed method was applied to a realistically sized sample: a 6 m offset-fed mesh antenna with a constraint on the half path-length best-fit tessellation error. After optimization, the converged solution nearly doubled the relative force tolerance without notably altering the tessellation error. Such a large improvement has significant practical implications: tolerance for manufacturing errors can be increased, and the permissible working environment can be expanded. The method of robust design presented in this chapter is simple and efficient, and it can be readily applied to actual scenarios.

Chapter 4
Effect and Design of Gravity Compensation System of Mesh Antennas on Deployment

4.1 Introduction

Large deployable structures, such as mesh antennas, are among the most valuable appendages of satellites; they are used for certain sophisticated tasks, such as high-resolution Earth observations and instant telecommunications. Structurally, this kind of antenna is primarily composed of a supporting truss and a flexible cable network arranged in a geometric paraboloidal shape to enable reflection of signals by the attached metallic mesh. The truss is usually a foldable mechanism with a single degree of freedom. The truss, driven by active motors or passive springs, unfolds the antenna from the launch ready stowed configuration to the working configuration in orbit. Failure of any portion of the unfolding mechanism is unacceptable as it may immediately result in malfunctions of the satellites. Therefore, accurate on-ground experimental verification of the deployment dynamics of mesh antennas is of great concern in the space industry.

One of the main problems encountered in using on-ground experiments to predict the in-orbit performances of antenna deployment is due to gravity. In-orbit deployments are performed under minor or even zero gravity, while there is always one standard gravity (g) on the Earth. Although various techniques compensating for the effects of gravity on deployment tests have been proposed, it has remained impossible to eliminate all of the effects of gravity [130]. Therefore, there is always some degree of difference between on-ground verification data and in-orbit flight data. These differences should be reduced to the extent possible by designing gravity compensation systems based on a proper and thorough understanding of the dynamical differences between on-ground deployment and in-orbit deployment.

© The Author(s), under exclusive license to Springer Nature Singapore Pte Ltd. 2026

K. Fu, *Design and Prediction of In-orbit Dynamic Behavior of Deployable Mesh Antennas*, Springer Theses, https://doi.org/10.1007/978-981-92-0140-2_4

4.1.1 Gravity Compensation of Ring Truss

In general, the gravity compensation system of the deployable mesh antenna usually consists of a fixed radial guide rail, pulleys moving along the guide rail, and suspension strings. Gravity compensation system of deployable mesh antenna Astromesh-I has been introduced in literature [106], in which adjacent two-bay gravity compensation design has been drawn in Fig. 4.1a. The suspension node of the suspension string is chosen as the midpoint of the longeron P, the other end of the suspension string Q connects with a slider, and the length of the suspension string remains unchanged during the unfolding process. The deployment of Astromesh-I is achieved by adjusting the length of diagonals. A simple geometric analysis shows that the height of each centroid G remains constant during deployment under the premise of approximate rigidity of the truss. From the point of view of energy, the gravitational potential energy of the ring truss is constant in the unfolding process, and the gravity compensation system does not do work on the ring truss. Therefore, the gravity compensation system ensures the energy consistency between the in-orbit and on-ground environment.

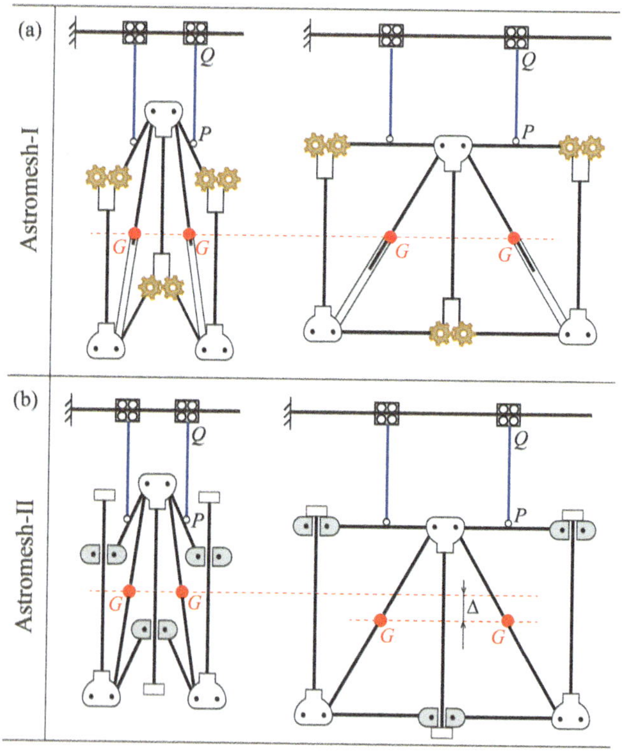

Fig. 4.1 Gravity compensation of the ring truss. (**a**) Astromesh-I; (**b**) Astromesh-II. Adapted with permission from [131]

For the gravity compensation of Astromesh-II, a simple design is to directly adopt the gravity compensation design of Astromesh-I, and the suspension node is still selected as the midpoint of the longeron P, as shown in Fig. 4.1b. Nevertheless, the geometric analysis shows that the centroid G of Astromesh-II decreases with the unfolding process, which indicates that the gravity of the ring truss does the positive work on the system during unfolding process. In other words, gravity contributes to the deployment of the antenna. Therefore, this design of gravity compensation system cannot achieve the expected effect of gravity compensation, even the on-ground experiment may mislead the prediction of in-orbit deployment behavior. Therefore, we should consider a new gravity compensation design to realize the gravity compensation of the ring truss.

4.1.2 Gravity Compensation of Webs

For the webs, the suspension system can be similarly designed by suspending several selected connection points on the cables with constant-length strings and/or constant-force strings. However, the ultra-flexibility of the webs makes finding the displacement relationship between the mass center of the webs and the suspension points complicated. As a result, the weight of the webs is not always counterbalanced by its suspension system during the unfolding process. It is expected that this imperfect offloading of the weight of the web network will have non-negligible effects on the deployment dynamics of the entire mesh antenna as the gravitational potential energy of the webs may be relevant. For example, in a 12.25 m mesh antenna, the mass of the webs accounts for about 1/5 of the total weight of the mesh antenna; in that case, if the webs are not offloaded at all during the unfolding process, as schemed in Fig. 4.2, the gravitational potential energy of the webs increases by about 20 times the increment amount of the elastic potential energy. The variation of gravitational potential energy can be easily understood because the webs go

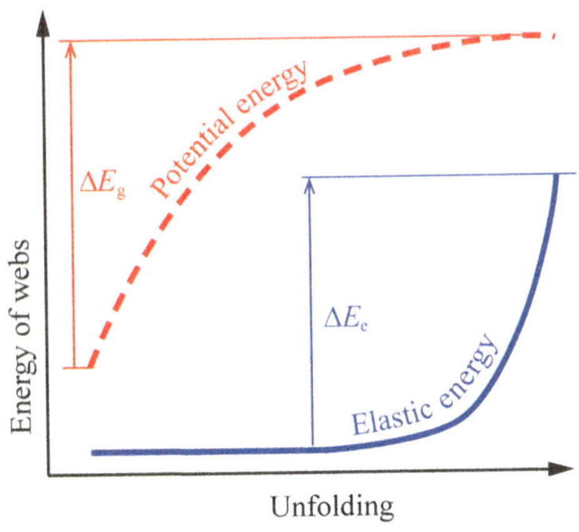

Fig. 4.2 Energy variation during unfolding process. Adapted with permission from [131]

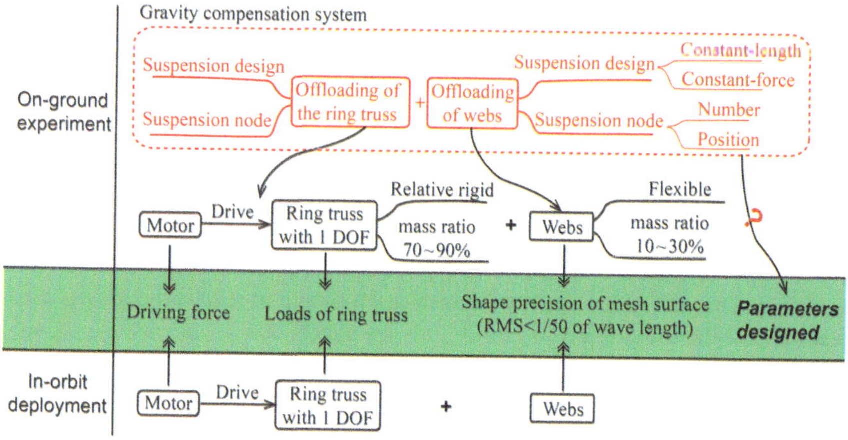

Fig. 4.3 Contents of this chapter. Adapted with permission from [131]

up during deployment. However, little research has been conducted on the gravity compensation systems for these webs. Up to now, how the web suspension system, such as the method, the number and the position of the suspension points, affects the overall deployment dynamics has remained unclear.

4.1.3 Contents of this Chapter

Based on a comprehensive understanding of the effect of gravity on deployment, a suitable gravity compensation system is designed for the ring truss and the webs, so that the on-ground experiment can predict the deployment behavior in orbit. The contents of this chapter are arranged as shown in Fig. 4.3.

4.2 Gravity Compensation Design of the Ring Truss

4.2.1 Problems of the Constant-Length Offloading

When the gravity compensation design of Astromesh-I is applied directly to the gravity compensation design of Astromesh-II, the geometric analysis of Fig. 4.4a shows that the center of gravity of the ring truss decreases gradually during the unfolding process, and the gravity contributes to deployment. The suspension node can be adjusted, such as the hinge point between the longeron and batten, or the hinge point among the longeron, batten and diagonal, as shown in Fig. 4.4b and c. It can be seen that when the suspension node is chosen as the hinge point between the longeron and batten, the height of the center of the single bay gradually decreases

Fig. 4.4 Design of constant-length offloading: (**a**) the midpoint of the longeron is selected as the suspension node; (**b**) the hinge point between the longeron and batten is selected as the suspension node; (**c**) the hinge point among the longeron, batten and diagonal is selected as the suspension node

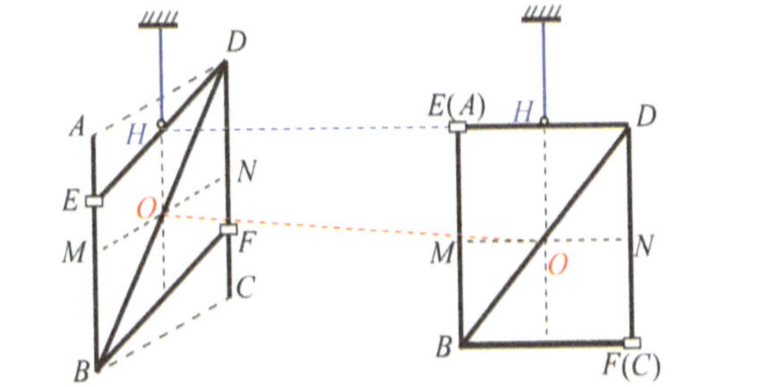

(a) the midpoint of the longeron is selected as the suspension node

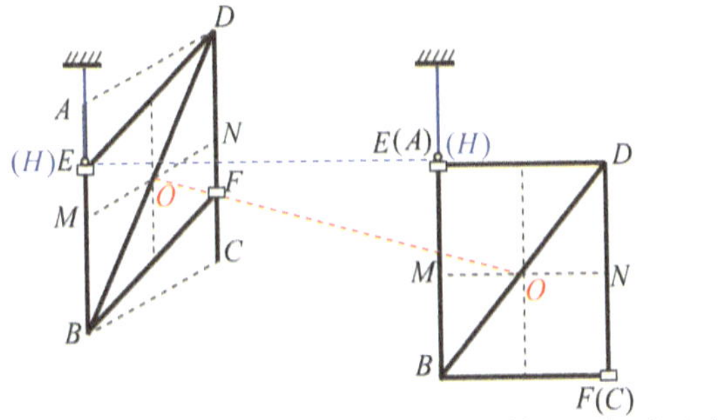

(b) the hinge point between the longeron and batten is selected as the suspension node

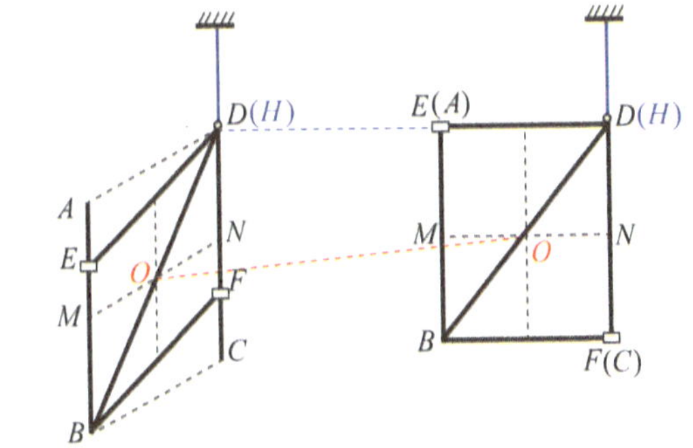

(c) the hinge point among the longeron, batten and diagonal is selected as the suspension node

during the unfolding process, resulting in gravity contributing to deployment. When the suspension node is chosen as the hinge point among the longeron, batten and diagonal, the height of the center of the single bay gradually increases during the unfolding process, resulting in gravity contributing to deployment. Therefore, the above plans cannot guarantee that the height of the center of the ring truss remains unchanged during the unfolding process.

As a matter of fact, a satisfactory design of gravity compensation for the ring truss is that the suspension node is selected as the midpoint of the diagonal O, but the suspension strings may interfere with the ring truss during the unfolding process, which is not feasible in practice. Therefore, it is necessary to design a new gravity compensation system for the ring truss.

4.2.2 Constant-Force Offloading Design of a Single Bay

There exist two principles to realize the gravity compensation of the ring truss. The first is to ensure that the height of the center of the ring truss remains constant during the unfolding process, such as the gravity compensation system of the ring truss for Astromesh-I. The second is to ensure the work done by gravity W_G is counteracted by the work done by gravity compensation systems W_F, i.e.

$$W_G + W_F = 0 \tag{4.1}$$

The above can also ensure that there exist no energetic exchange between the ring truss and the external environment.

The constant-force offloading of the ring truss is based on the second principle of gravity compensation as mentioned above. The constant force along the vertical upward direction is applied to the ring truss, and the appropriate suspension node is selected to achieve the complete gravity compensation. For the sake of simplicity, a single bay of Astromesh-II is chosen as the research object to conduct the gravity compensation.

The top nodes A and D are selected as the suspension node, as shown in Fig. 4.5. The length of the left suspension string AK is a constant, in which one end of the suspension string is connected at the suspension node A, the other end is fixed at K of the guide rail. One end of the right suspension string is fixed on the top of the right batten D, and the other end of the right suspension string goes through the pulley H and is fixed with a weight m_{R1}. The pulley H is fixed on a slider along the horizontal guide rail. The tension of the cable DH keeps constant, resulting in a constant-force offloading on the top node D. It can be seen that the height of the mass center O decreases during the unfolding process, and the appropriate value of the constant force should be determined to counteract the work done by the gravity.

The mass of each component of the single bay is labeled in Fig. 4.5, and the tension of cable DH is denoted as F_1. The displacement of the top node of the right batten is denoted as Δ, which is relative to the fully deployed configuration. Geometrically, the displacement of the center of the single bay can be expressed as $\Delta/2$. In practice, there is a large mass deviation of translational joints due to the geometric difference, while the mass deviation among the longerons can be negligible. In order to ensure the mass center of adjacent two translational joints coincides with the geometric center of the single bay, it is necessary to balance the mass of the translational joints. Given the gravitational acceleration g, if the deployed state is considered as the initial state, then the gravitational work done in the folding process can be calculated as

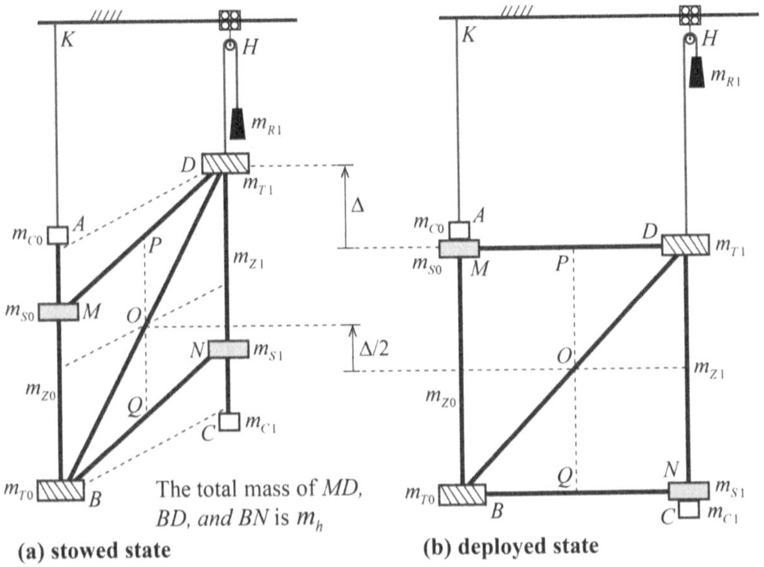

Fig. 4.5 Constant-force offloading of single bay: (**a**) stowed state; (**b**) deployed state

$$W_G = -m_h g \cdot \frac{\Delta}{2} - \left(m_{T1} + m_{Z1} + m_{C1} + m_S\right) g\Delta \tag{4.2}$$

The work done by the offloading constant force can be expressed as

$$W_F = F_1 \Delta \tag{4.3}$$

By substituting Eqs. (4.2)–(4.3) into (4.1), the offloading constant force can be calculated as

$$F_1 = \left(m_S + \frac{m_h}{2} + m_{T1} + m_{Z1} + m_{C1}\right) g \tag{4.4}$$

The tension of the cable *AK* can be expressed as

$$F_0 = \left(m_S + \frac{m_h}{2} + m_{T0} + m_{Z0} + m_{C0}\right) g \tag{4.5}$$

There are many ways to realize constant-force offloading such as a cable-pulley system with a weight and a constant-force spring. If the cable-pulley system with a weight is adopted to realize the constant-force offloading, the mass of the weight m_{R1} can be expressed as

$$m_{R1} = m_S + \frac{m_h}{2} + m_{T1} + m_{Z1} + m_{C1} \tag{4.6}$$

Correspondingly, the suspension string AK can be replaced with the cable-pulley system with a weight, and the mass of the weight m_{R0} can be expressed as

$$m_{R0} = m_S + \frac{m_h}{2} + m_{T0} + m_{Z0} + m_{C0} \tag{4.7}$$

4.2.3 Constant-Force Offloading Design of the Ring Truss

The previous section presented the constant-force offloading of a single bay, here the constant-force offloading will be extended to the ring truss which is composed of multiple bays. Figure 4.6 shows the schematic diagram of the constant-force offloading of the ring truss. The motors are located at the side AB. One end of the cable AK is fixed on the top of the batten, and the other is fixed on the guide rail. Constant forces are applied on the other tops of the battens, in which the cable-pulley system with a weight is adopted to realize the constant force as shown in Fig. 4.6. Each pulley is fixed on a slider along the horizontal guide rail.

Taking the first to third bays as shown in Fig. 4.6b as an example, the calculation of balancing the mass of each bay and the calculation of constant force will be illustrated. Since the first and second bay share the batten CD, the fixed connections C and D, and the translational joint N, it is reduced to half of their respective mass on each single bay. Due to the mass deviation of translational joints in practice, it may be considered that the mass of translational joints is related as follows

$$m_{S1} > m_{S0}, m_{S1} > m_{S2}, m_{S3} > m_{S2} \tag{4.8}$$

In order to ensure the mass center of the two translational joints locates at the geometric center, the mass of translational joints of each span should be balanced. The right half of the translational joint M needs to be added with a counterweight Δ_0^r, and the left half and right half of the translational joint P need to be added with a balancing mass Δ_2^l and Δ_2^r respectively. The following requirements should be satisfied

$$\begin{cases} \frac{1}{2}m_{S0} + \Delta_0^r = \frac{1}{2}m_{S1} \\ \frac{1}{2}m_{S2} + \Delta_2^l = \frac{1}{2}m_{S1} \\ \frac{1}{2}m_{S2} + \Delta_2^r = \frac{1}{2}m_{S3} \end{cases} \tag{4.9}$$

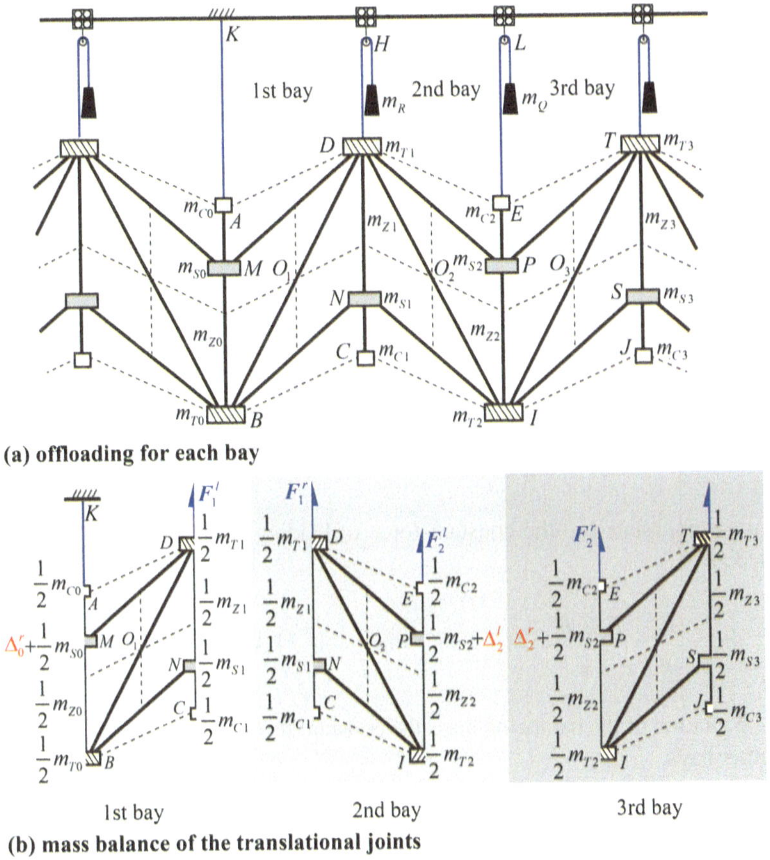

Fig. 4.6 The constant-force offloading of the ring truss: (**a**) offloading for each bay; (**b**) mass balance of the translational joints

Therefore, the balancing mass which should be added to the translational joint P can be obtained as

$$\Delta_2 = \Delta_2^l + \Delta_2^r \tag{4.10}$$

According to the Eqs. (4.4) and (4.5), the constant forces applied on the each bay can be expressed as

$$\begin{cases} F_1^l = \frac{1}{2}\left(m_{S1} + m_{h1} + m_{T1} + m_{Z1} + m_{C1}\right)g \\ F_1^r = \frac{1}{2}\left(m_{S1} + m_{h2} + m_{T1} + m_{Z1} + m_{C1}\right)g \\ F_2^l = \frac{1}{2}\left(m_{S2} + m_{h2} + m_{T2} + m_{Z2} + m_{C2} + 2\Delta_2^l\right)g \\ F_2^r = \frac{1}{2}\left(m_{S2} + m_{h3} + m_{T2} + m_{Z2} + m_{C2} + 2\Delta_2^r\right)g \end{cases} \tag{4.11}$$

The mass m_R related to the constant-force offloading of the batten *CD* can be calculated as

$$m_R = \frac{F_1}{g} = \frac{F_1^l + F_1^r}{g} = m_{S1} + m_{T1} + m_{Z1} + m_{C1} + \frac{1}{2}\left(m_{h1} + m_{h2}\right) \tag{4.12}$$

The mass m_Q related to the constant-force offloading of the batten *EI* can be calculated as

$$m_Q = \frac{F_2}{g} = \frac{F_2^l + F_2^r}{g} = m_{S2} + m_{T2} + m_{Z2} + m_{C2} + \Delta_2^l + \Delta_2^r + \frac{1}{2}\left(m_{h2} + m_{h3}\right) \tag{4.13}$$

The calculation of the balancing mass the constant force is the same as the above for the other bays.

4.3 Constant-Force Offloading Experiment

4.3.1 *Constant-Force Offloading Experiment of a Single Bay*

In order to examine the effect of constant-force offloading, an experiment was carried out on a single bay based on our lab's in-house model. The length and mass of the bars and joints can be found in Table 4.1. According to the requirements of the constant force in previous section, the mass center of adjacent translational joints should locate at the geometric center of the single bay. Considering that the difference of the mass between the adjacent translational joints, the mass of the adjacent translational joints should be balanced. In detail, $\Delta m = 404.4 - 328.2 = 76.2$ g should be added to the left translational joints. The difference of the mass between the upper and lower longerons is less than 1 g which can be ignored. And the mass of the left and right batten is not required to be the same. According to the Eqs. (4.4) and (4.5), the left constant force can be calculated as $F_L = 11.32$ N, and the right constant force can be calculated as $F_R = 11.15$ N.

Table 4.1 Parameters of a single bay

	Parameter	Value
Bar	Length of longeron	1.0 m
	Mass of upper longeron	135.6 g
	Mass of lower longeron	134.8 g
	Length of batten	1.4 m
	Mass of left batten	382.4 g
	Mass of right batten	389.5 g
Joint	Mass of left fixed connection	55.1 g
	Mass of left translational joint	328.2 g
	Mass of left T-shaped connection	290.1 g
	Mass of right fixed connection	75.3 g
	Mass of right translational joint	404.4 g
	Mass of right T-shaped connection	245.5 g

In the experiment, a cable-pulley system with a sandbag is adopted to realize the constant-force offloading. Compared with the constant-force spring, this method is simple and convenient for adjusting the force amplitude, especially suitable for the concept verification and iteration before final determination.

The constant-force offloading was conducted based on a single bay and the experimental results were as shown in Fig. 4.7:

1. When the theoretical values of constant force were chosen as $m_L = 1.032$ kg and $m_R = 1.215$ kg, the single bay can remain stable, indicating that the influence of gravity can be nearly eliminated.
2. When the left constant force was less than the theoretical value and the right constant force was larger than the theoretical value as $m_L = 1.032$ kg and $m_R = 1.215$ kg, the experimental results showed that the single bay would gradually fold, indicating that the influence of gravity cannot be eliminated and the gravity contributes to folding.
3. When the left constant force was larger than the theoretical value and the right constant force was less than the theoretical value as $m_L = 1.232$ kg and $m_R = 1.015$ kg, the experimental results showed that the single bay would gradually folded, indicating that the influence of gravity cannot be eliminated and the gravity contributes to unfolding.

The above experimental results are consistent with theoretical expectations. It can be seen that the constant force can eliminate the influence of gravity on deployment when it meets the requirements given by Eqs. (4.4) and (4.5).

4.3.2 *Constant-Force Offloading Experiment of the Ring Truss*

In order to further validate the constant-force offloading design, a gravity compensation experiment of the ring truss was conducted on a six-bay ring truss with 2 m diameter. The number of battens was labeled in the top right corner of Fig. 4.8, and

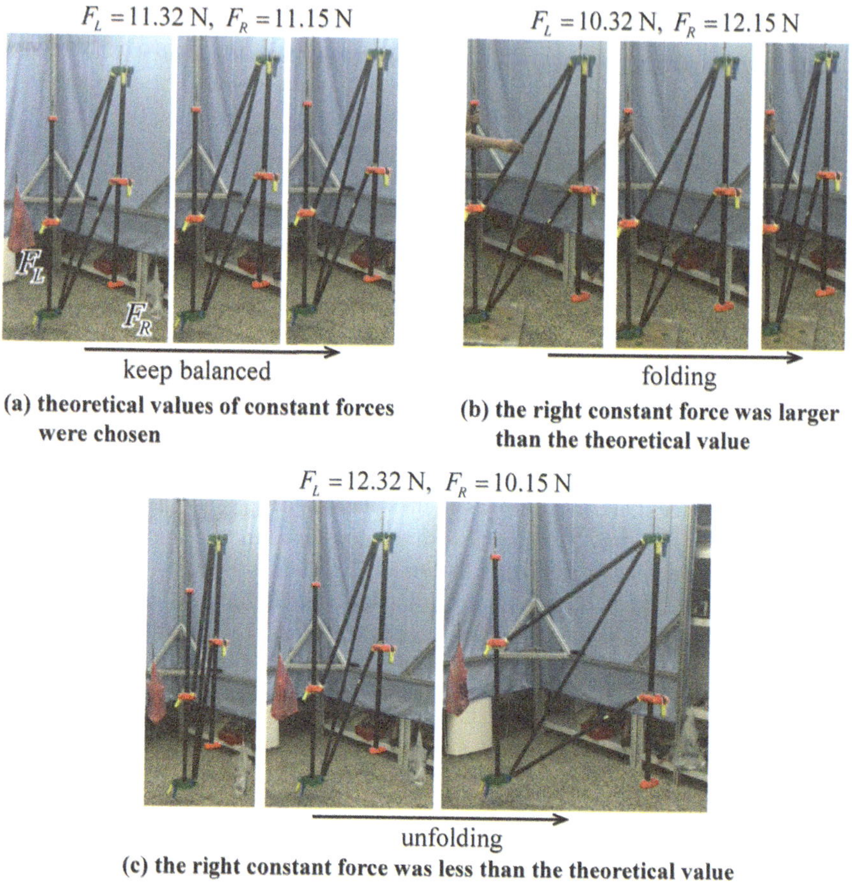

Fig. 4.7 Constant-force offloading experiment of a single bay: (**a**) theoretical values of constant forces were chosen; (**b**) the right constant force was larger than the theoretical value; (**c**) the right constant force was less than the theoretical value

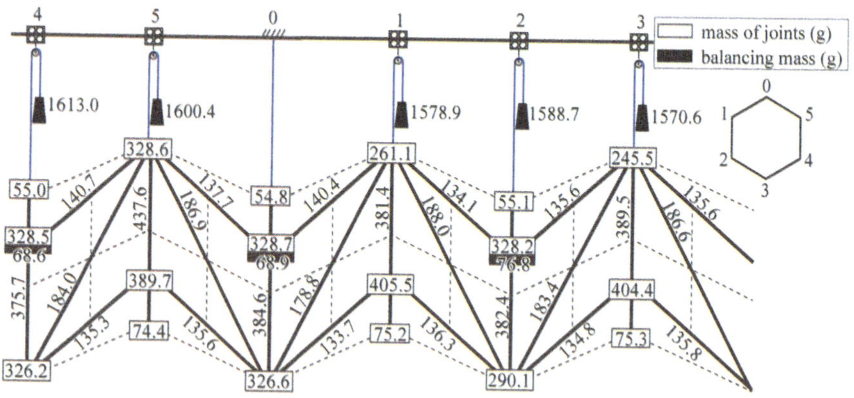

Fig. 4.8 Mass balance of the ring truss

the mass of each component was marked in Fig. 4.8. Firstly, according to Eqs. (4.9) and (4.10), the masses of translational joints were balanced to ensure that the mass center of the two adjacent translational joints is located at the geometric center of each bay. Then, Eqs. (4.12) and (4.13) were used to calculate the constant force. The calculation results of the balancing masses and counterweights were given in Fig. 4.8.

The constant-force offloading experiment of the ring truss was shown in Fig. 4.9, in which whether the ring truss under the constant-force offloading can maintain its configuration or not was examined. Experimental results showed that the ring truss was able to maintain its configuration including the stowed state, the deployed state and the transition state, which indicates that the influence of gravity on deployment can be eliminated by constant-force offloading.

4.4 Effect of Gravity of Webs on Deployment Dynamics

For a 12.25 m mesh antenna, the mass of the webs accounts for about 1/5 of the total weight of the mesh antenna; in that case, if the webs are not offloaded at all during the unfolding process, the gravitational potential energy of the webs increases by about 20 times the increment amount of the elastic potential energy. The effect of gravity of webs on deployment dynamics cannot be ignored. This section will mainly discuss the effect of gravity of webs on deployment dynamics.

4.4.1 Modeling a Mesh Antenna Utilizing a Flexible Multibody Approach

Flexible multibody simulation models of mesh antennas were built to study the deployment dynamics of such antennas both on-ground and in-orbit. A 12.25 m Astromesh-II antenna, which includes a deployable ring truss with 30 bays, was used as the basis for the model. The studied antenna is circular and front-fed, for simplicity of analysis; however, the adopted method and the obtained results are also suitable for elliptic and offset-fed antennas. Figure 4.10 shows the overall structure of the studied antenna and the local structure of a two-bay truss. Two adjacent translational joints along the battens are synchronized by synchronization cables, shown as blue solid lines. In addition, two driving cables, shown as red solid lines, are used to drive the translational connections and unfold the mesh antenna. The ring truss is deployed by reeling in the two driving cables, whose ends are controlled by four motors. Two motors are fixed at the upward batten of the first bay while the other two are fixed at the downward batten. One design of the driving cable is shown in Fig. 4.10, in which one single cable passes through a series of pulleys fixed on the connections (the pulley is not shown). When the driving cables

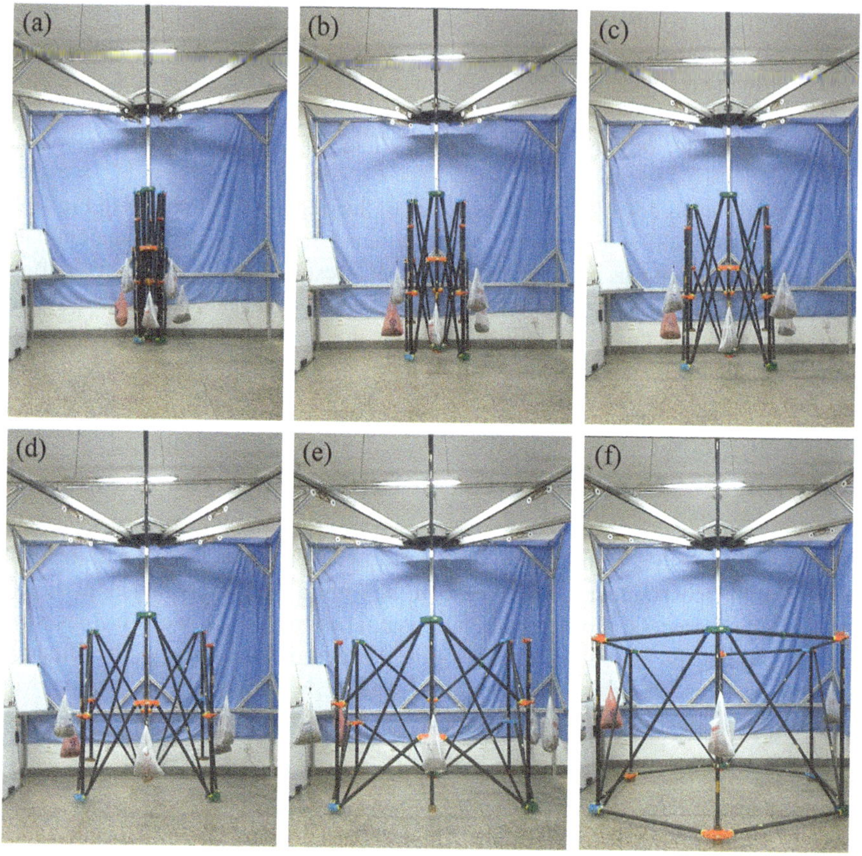

Fig. 4.9 Constant-force offloading experiment of the ring truss: (**a**) stowed state; (**b**–**e**) transition state; (**f**) deployed state

are reeled in, the translational connections will move towards the fixed connections, and therefore the ring truss is approaching deployment.

The entire simulation model was built and solved in our lab's in-house multibody code. As shown in Fig. 4.10, the majority of the components in the mesh antenna were modeled with conventional modeling techniques developed under a flexible multibody framework, which classifies various components as elements, constraints, and forces. In brief, the ring truss assembly was modeled by periodically repeating a two-bay model in which all of the connections, including T-shaped connections, fixed connections, and translational connections, were regarded as rigid bodies; all of the longerons, battens, and diagonals were meshed with geometrically exact beam elements. All the mechanical joints, including revolute, fixed, and translational joints, were formulated by constraint equations. The latch mechanisms for the translational joints were simplified to nonlinear forces only since the translational joints have already constrained the attitudes of the translational connections.

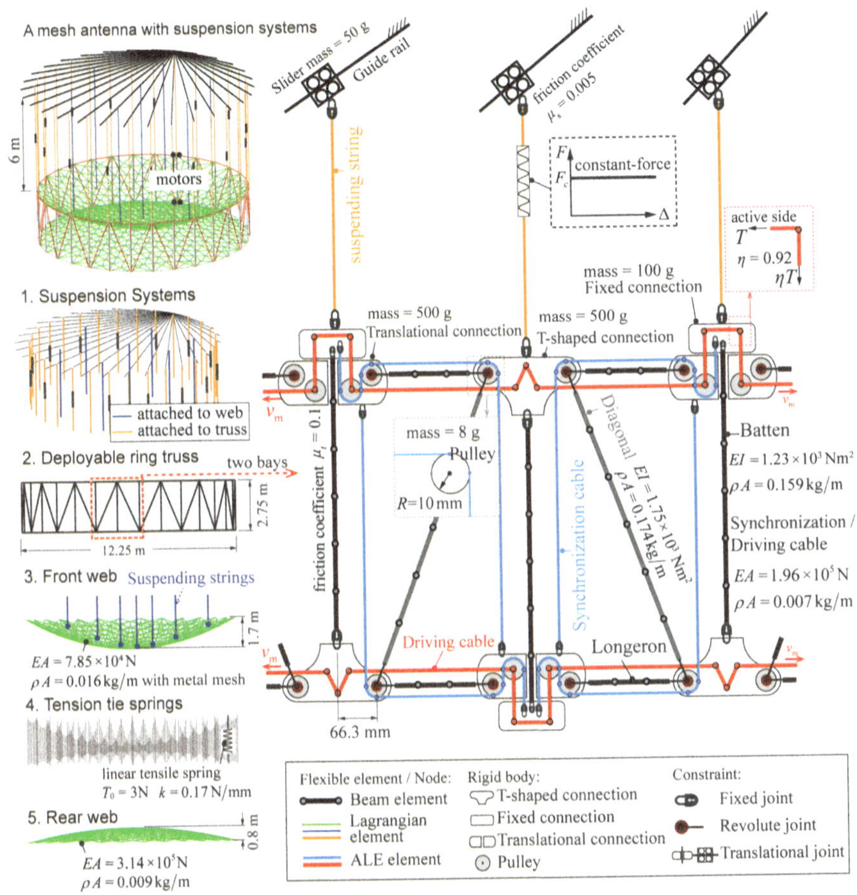

Fig. 4.10 Flexible multibody model of a 30-bay mesh antenna with gravity compensation systems. Adapted with permission from [131]

Cables in the webs and the suspension systems were meshed with cable elements. The tie springs between the front web and the rear web were modeled with linear tensile forces, and the constant-force springs in the suspension systems were replaced with constant tensile forces.

Figure 4.10 also offers detailed geometric and mechanical parameters of the antenna, such as bending stiffnesses *EI* of the beam elements, tensile stiffnesses *EA* of the cable elements, linear densities A of the beam and cable elements, and pretension force T_0 and stiffness *k* of the tie springs. In this model, the mass of the metal mesh was uniformly added to the front web. For simplicity, the masses of the constant-force springs were added to the sliders on the guide rails. As a result, the constant-force springs will not add redundant gravity which needs to be unloaded while the inertial effect of the constant-force springs is still considered. The friction force within any translational joint, such as between a slider and a guide rail, and between a batten and a translational connection, was calculated by multiplying the constraint force of the joint and the corresponding friction coefficient μ; the direction is opposite to the relative velocity of the two sliding bodies. Here a stick-slip friction model is adopted to characterize the translational friction, which enables to generate a frictional force when the relative velocity reaches zero.

Although the above mentioned components may be modeled in a straightforward manner with well developed techniques, there is a notable difficulty in building an efficient model of the synchronization and driving cables that takes into account their contact interactions with the pulleys. The traditional modeling method meshes the cables with the cable elements and calculates the contact forces and friction forces between the cable elements and the pulley. Because the mesh size of the elements should be much smaller than the pulley radius, this method will involve a huge number of cable elements and detections of potential contact events, resulting in extremely low calculation efficiency. This low-efficiency problem used to make it nearly impossible to finish a full-scale deployment simulation of a large mesh antenna. Fortunately, this problem has been recently solved by introducing a cable element of variable length based on the arbitrary Lagrangian–Eulerian (ALE) formulation that uses a generalized force to account for the cable tension loss between the two sides of the pulley caused by friction and adopting constraint equations to dynamically locate the contact border of the pulley and the cable.

Using this multibody approach, a specific simulation model for a 30-bay, 12.25 m mesh antenna was built, and the relevant modeling parameters were included in Fig. 4.10 to provide more complete information. Besides, the reel-in velocity v_m decreases when the driving force T_m increases with the relationship $v_m = 20 - 0.01T_m$. The unit of v_m is millimeters per second and the unit of T_m is Newtons.

For the mesh antenna model, the system contains three main kinds of friction: friction within the driving cables, described by the transmission efficiency η; friction within the translational joints, described by the translational friction coefficient μ; and friction within the sliders along the guide rails, described by the sliding friction coefficient μ_s. Here for the current study, η is set to be 0.92, μ is set to be 0.10, and μ_s is set to be 0.005. In practice, these parameters are caused by the complex contact details and should be adjusted according to the experiment. For example, η is a function of many physical conditions, such as the working temperature and the cable tension. However, it is difficult to obtain a theoretical relationship between η and these physical parameters, which are caused by the complex contact details within bearing or between the pulley and the shaft.

Due to the efficiency of this modeling approach, the entire unfolding process could be calculated in around 10 h with a four-core personal computer (Intel Core i7–4770 3.4 GHz). Then, both in-orbit and onground deployment processes were systematically simulated and analyzed.

4.4.2 In-Orbit Deployment Dynamics

The in-orbit deployment dynamics of the mesh antenna were first analyzed to serve as a reference case for further on-ground studies. The simulated time history of the driving force is plotted by the red solid line in Fig. 4.11a. The entire unfolding process can be roughly separated into two stages, in which the Stage-I is referred to as the early deployment period when the webs are slack, while the Stage-II refers to as the late deployment period when the webs are tensioned. For the current simulation, it can be found that the demarcation of the two stages locates at around $t = 600$ s. In Stage-I from 0 s to 600 s, the driving force remains small, not exceeding 10 N. However, in Stage-II from 600 s onward, the force sharply increases. In Stage-I, the early deployment period from 0 s to 600 s, the driving force remains small, not exceeding 10 N. However, in Stage-II, the late deployment period from 600 s onward, the force sharply increases.

For better understanding of the deployment behavior, a semi-analytical analysis of the driving force was conducted from an energetic viewpoint. Theoretically, the energy inputted by the motors is partially transferred into the potential energy (PE) of the mesh antenna and partially dissipated by frictions, which can be expressed as

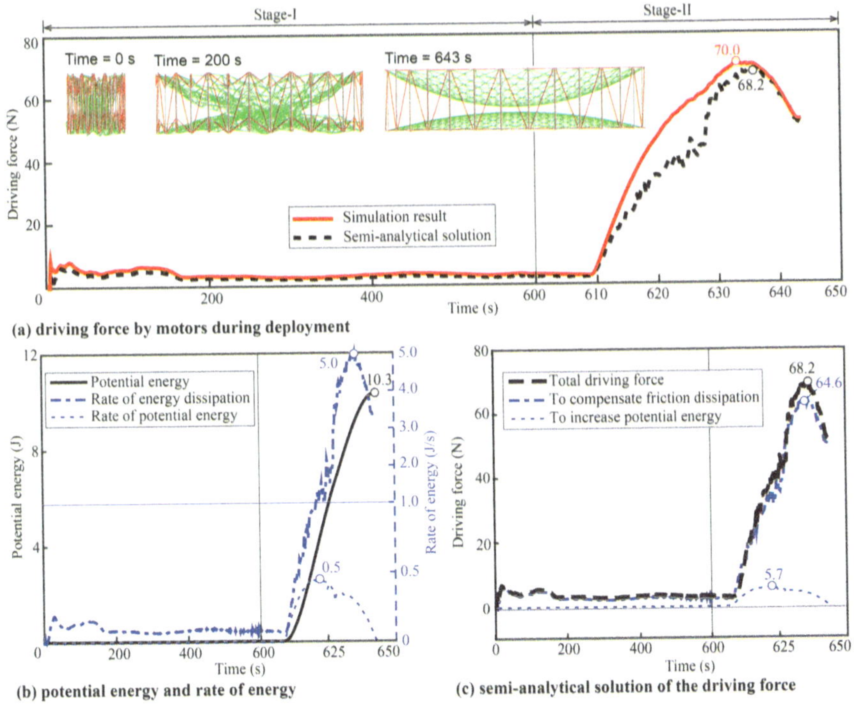

Fig. 4.11 In-orbit deployment. (**a**) driving forces by motors during deployment; (**b**) potential energy and rate of energy; (**c**) semi-analytical solution of the driving force. Adapted with permission from [131]

$$\dot{W}_{\text{motors}} = \dot{W}_{\text{PE}} + \dot{W}_{\text{friction}} \tag{4.14}$$

where $\dot{W}$ denotes the time-varying rate of the energy, and

$$\dot{W}_{\text{motors}} = 4T_{\text{m}}v_{\text{m}} \tag{4.15}$$

due to the symmetry of the structure. Because the driving force is small (<80 N), it can be obtained that $v_m \approx 20mm/s$ according to the expression of the reel-in velocity. Substituting Eq. (4.15) into (4.14) gives

$$T_{\text{m}} \approx \frac{1}{80}\left(\dot{W}_{\text{PE}} + \dot{W}_{\text{friction}}\right) \tag{4.16}$$

This equation means that the driving forces need to simultaneously fulfil the requirement of increasing the potential energy and overcoming the friction forces.

The potential energy of the entire mesh antenna during the deployment process was obtained through simulation and plotted by the black solid line in Fig. 4.11b. It is small during the early deployment stage and grows sharply during the late deployment stage, as the webs are tensioned only when the antenna is close to its fully deployed configuration. As a result, $\dot{W}_{\text{PE}}$, plotted by the thin blue dotted line in Fig. 4.11b, is notable only within late stage and reaches its maximum shortly before fully deployed configuration. In addition, $\dot{W}_{\text{friction}}$ was obtained by summing up the friction works within all of the cable-pulley systems and translational joints, as plotted by the thick blue dashed-dot line in Fig. 4.11b. Obviously, $\dot{W}_{\text{friction}}$ is well correlated with $\dot{W}_{\text{PE}}$ as the increase in potential energy is accompanied by the increase in the friction forces.

According to Eq. (4.16), summing up $\dot{W}_{\text{friction}}/80$ and $\dot{W}_{\text{PE}}/80$ gives the driving force, obtained in a semi-analytical way, as shown by the thick black dashed line in Fig. 4.11a and c. It matches well with the simulated driving force, shown by the red solid line.

4.4.3 On-Ground Deployment Dynamics Without Web Suspension Systems

Before investigating the web suspension system, this section simulates the on-ground deployment dynamics of the mesh antenna without any web suspension system but with the truss suspension system. The results are compared with the results of in-orbit deployment to reveal some insights on the effects of gravity. Moreover, the on-ground deployment without a web suspension system also serves as a reference case for the following sections.

Figure 4.12a shows the driving forces of the motors during the unfolding process. The on-ground and in-orbit cases are shown in red dashed line and black solid

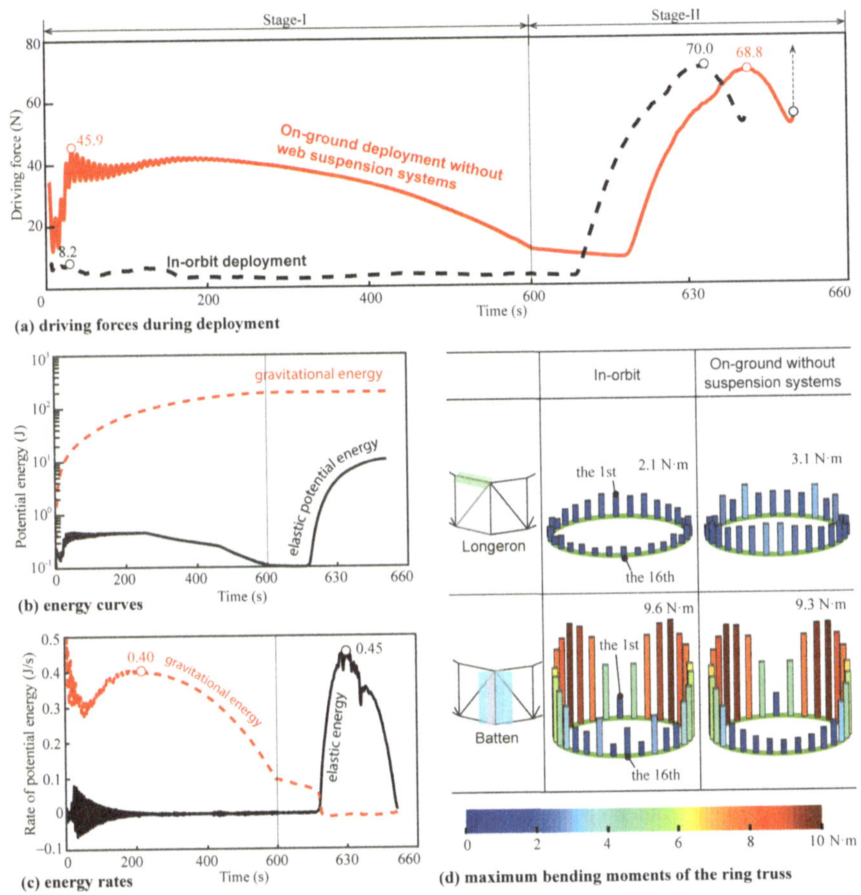

Fig. 4.12 On-ground deployment without web suspension systems. (**a**) driving forces during deployment; (**b**) energy cueves; (**c**) energy rates; (**d**) maximum bending moments of the ring truss. Adapted with permission from [131]

line, respectively. In Stage-I, the early period of deployment, the maximum force of the on-ground case is much larger than that of the in-orbit case. However, in Stage-II, the late period of deployment, the force curves of the two cases share the similar shapes. They both experience a sharp increase with almost identical peak force values of approximately 70 N.

The difference in the behavior of the driving forces during the early and late unfolding stages can be roughly explained from an energetic point of view. Theoretically, the work done by the driving forces is partially transferred into the elastic and gravitational potential energies of the webs and partially dissipated by friction. Because the friction forces are nearly proportional to the driving forces, it is expected that the trends of the driving forces will have similar shapes to the rate changes of the elastic and potential energies. The potential energies and their changing rates are plotted in Fig. 4.12b and c. Overall, the gravitational potential energy is approximately one to two orders of magnitude larger than the elastic energy. However, the variation in the gravitational energy occurs primarily in the very early unfolding period and then gradually decreases as deployment commences. Correspondingly, the driving force is larger in the on-ground case than in the in-orbit case, and there exists one peak force at Stage-I for the on-ground case. Additionally, at the late stage of unfolding (Stage-II), the rate of change of the gravitational energy slows to nearly zero, while the elastic energy of the web networks increases sharply similar to the in-orbit case. As a result, the driving force during the late deployment is dominated by the deformation of the webs instead of by the gravitational energy of the webs. Therefore, the driving forces of the in-orbit and on-ground cases have nearly identical peak values.

In addition to monitoring the driving forces of the motors, the maximum bending moments of the truss members during the entire unfolding process, were also monitored as shown in Fig. 4.12d, including all of the longerons and battens. For the longerons, no external loads are applied at the two rotation-jointed ends, resulting in little bending moments in the parallelogram plane of each bay. The simulation results also support this conjecture, as the maximum moment in the parallelogram plane is less than 0.01 N·m. However, the bending moments in the cross plane of the cylindrical ring truss are not close to zero. For the on-ground case, the maximum bending moment is 3.1 N·m, about 50% larger than the moment for the in-orbit case, 2.1 N·m. For the battens, the bending moments in both the radial and tangential planes of the cylindrical ring truss were measured. The results show no obvious differences between the on-ground and in-orbit cases, with peak values 9.6 N·m and 9.3 N·m, respectively.

In summary, the gravity of webs has non-negligible effects on the deployment dynamics of mesh antennas, with a relatively large effect on the driving force and a relatively small effect on the bending moments of the truss members. The driving forces for the on-ground case are larger than those for the in-orbit case during the early deployment stage (Stage-I), but the two cases have almost the same peak value at the late deployment stage (Stage-II). For the bending moments of the truss members, the main difference occurs on the longerons.

4.5 On-Ground Deployment Dynamics with Web Suspension Systems

4.5.1 Two Designs of Web Suspension Systems

The results described in the previous section show that, if the webs of the mesh antenna are not offloaded, on-ground tests cannot accurately predict the in-orbit behaviors, especially in terms of the driving forces of the motors at the early deployment stage. To obtain a more accurate prediction, two kinds of web suspension systems were designed and evaluated for compensation of the weight of the webs.

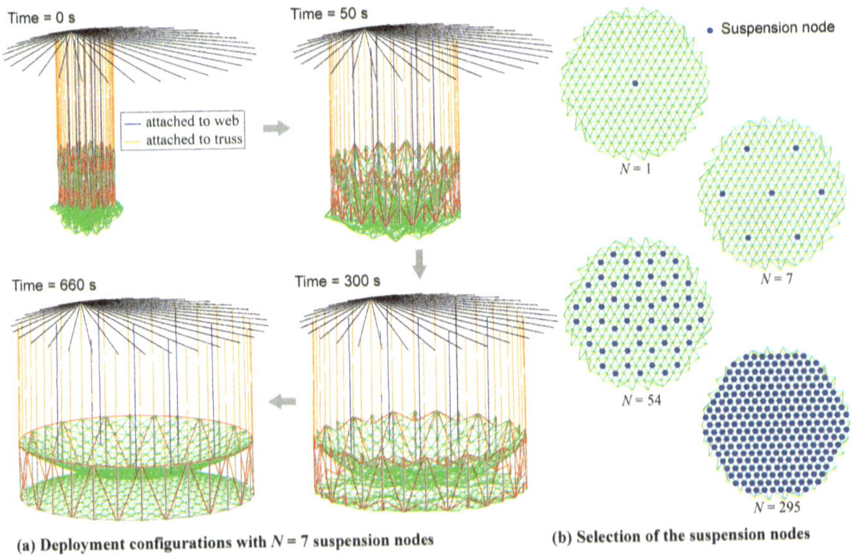

Fig. 4.13 Constant-length offloading. (**a**) deployment configurations with $N = 7$ suspension nodes; (**b**) selection of the suspension nodes. Adapted with permission from [131]

The first suspension system, referred to as the constant-length-design, suspends several points of the front web with constant-length suspension strings. These constant-length strings are plotted as blue solid lines in Fig. 4.13a. The second suspension system, referred to as the constant force-design, uses constant-force suspension strings to suspend the webs. All of the constant-force strings provide the same constant suspension force, and their sum value equals the total weight of the webs.

In designing the web suspension system, it is natural to ask which and how many nodes of the front web should be selected as the suspension nodes. The mesh antenna studied here has 295 cable nodes in total. As shown in Fig. 4.13b, four different numbers and arrangements of suspension nodes are studied, where N is 1, 7, 54 and 295, respectively; $N = 1$ means that only the central node of the front web is suspended and $N = 295$ means that all of the nodes are suspended. The performance of each web suspension system was evaluated in terms of both the driving forces and the bending moments and compared with the in-orbit case and the on-ground case without a web suspension system.

4.5.2 *Evaluation of Gravity Compensation*

Figure 4.14a–f compare the results of the constant-length-design and the constant-force-design. Two interesting and heuristic conclusions can be drawn from these results. One is that the performance of the web suspension system may not increase monotonically with the increase in the number of suspension nodes, in terms of

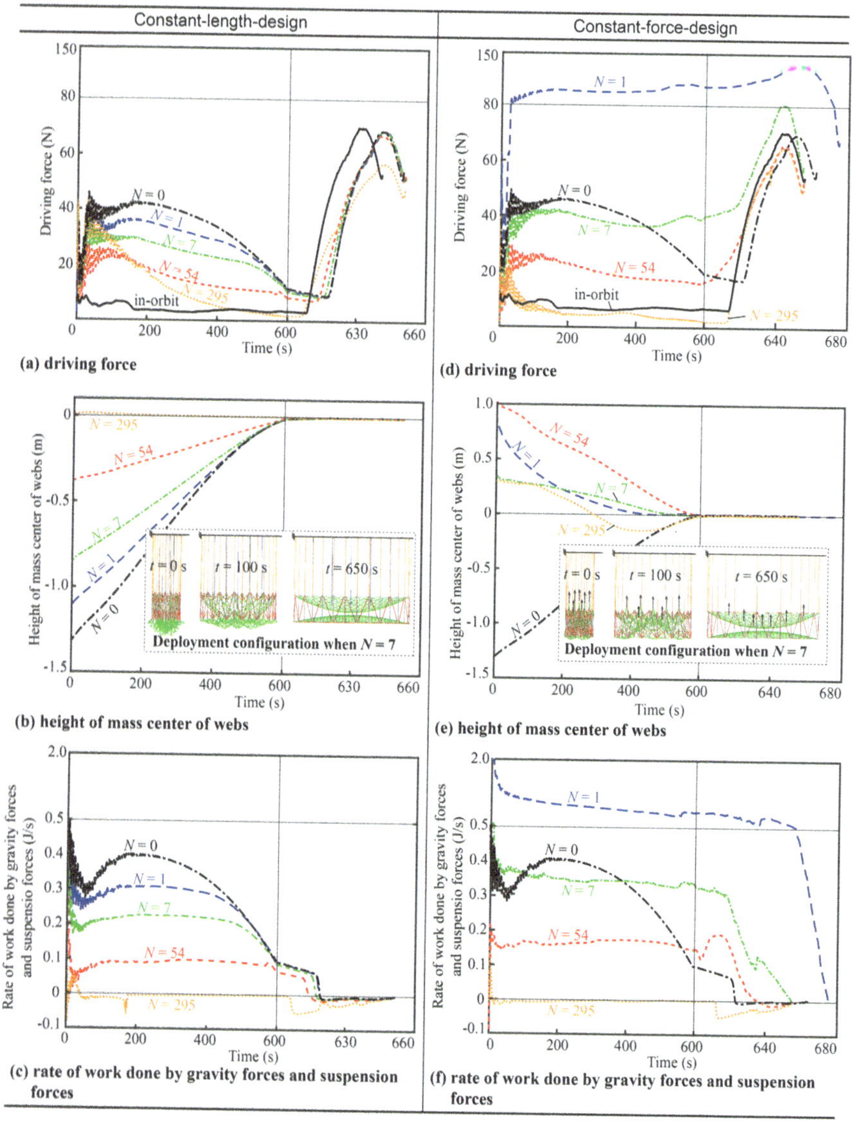

Fig. 4.14 Effects of web suspension systems on deployment. (**a, d**) diving force; (**b, e**) height of mass center of webs; (**c, f**) rate of work done by gravity forces and suspension forces. Adapted with permission from [131]

driving forces shown in Fig. 4.14a and d. For example, the constant-length-design with $N = 295$ performs worse than the other designs at the late deployment stage (Stage-II), while the constant-force-design with $N = 1$ performs even worse than $N = 0$ across the entire unfolding process. In other words, only systems utilizing specific numbers of suspension nodes have positive results in terms of reducing the difference between the on-ground verification and the in-orbit deployment. The other conclusion is that the constant-length-design performs better than the constant-force-design as the latter design always changes the peak force at the late unfolding stage.

These results can also be roughly explained from an energetic viewpoint. Ideally, the purpose of the gravity compensation system for the webs is to make the total work done by the gravitational forces and the suspension forces balance to zero.

For the constant-length-design, almost no work is done by the suspension system because the height variation of each suspension node is close to zero during deployment. However, the height of the mass center of the webs changes during deployment, as shown in Fig. 4.14b. Figure 4.14c plots the rates of total work done by both gravity forces and the suspension forces, whose values should be balanced to zero if the webs are fully offloaded. However, they are not zero. Moreover, the nonzero value implies that, at the early unfolding stage, the deviation of the driving force between the in-orbit and on-ground deployments monotonically decreases with the increasing number of suspension nodes. Additionally, when all of the web nodes are suspended, the mass center of the webs decreases at the late unfolding stage, making the driving forces smaller than those experienced in-orbit. Therefore, a moderate number of suspension nodes is a better choice.

For the constant-force-design, the work done by the suspension system is not only nonzero but also remarkable due to the notable height changes of the suspension nodes. The total rates of work done by both gravity forces and the suspension forces are plotted in Fig. 4.14f. Their values are generally much larger than those for the constant-length-design, plotted in Fig. 4.14c. That is why the constant-force-design generally performs worse than the constant-length-design.

Finally, an overall comparison between the constant-length-design and constant-force-design is provided in Fig. 4.15. It takes into account both the driving forces and the maximum bending moments. For the constant-length-design, it can be concluded that the compensation performance relies to a large extent on the number of suspension nodes. Too few ($N = 1$) or too many ($N = 295$) nodes will weaken the offloading performance. The constant-force-design shows terrible performance when $N = 1$, while better compensation can be achieved when 7 or 54 suspension nodes are used. Regardless of whether the constant-length-design or the constant-force-design is used, a moderate number of suspension nodes should be chosen to reduce the height variation of the mass center of webs as well as to reduce the additional inertial effects caused by the suspension system itself.

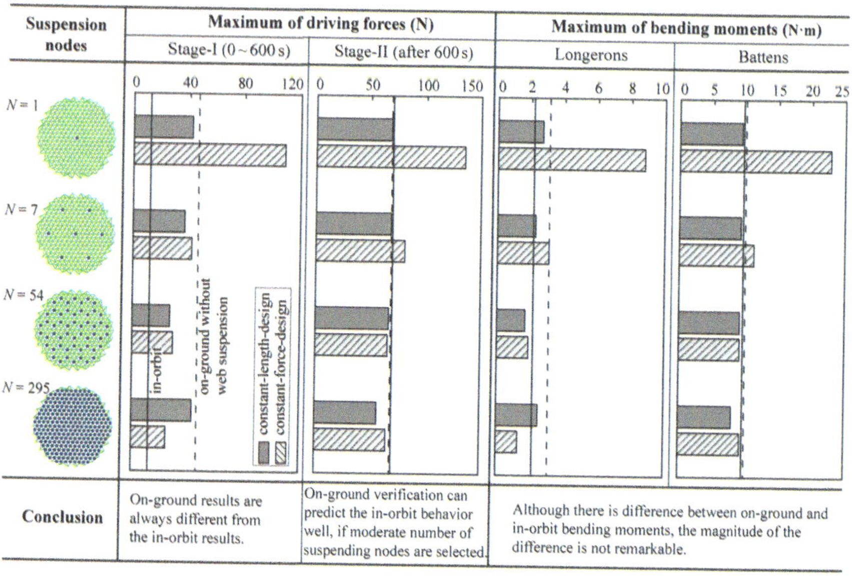

Fig. 4.15 Overall comparison between the constant-length-design and the constant-force-design. Adapted with permission from [131]

4.6 Summary

In-orbit prediction of deployment dynamics of mesh antennas is severely hampered by the effects of gravity, which cannot be completely eliminated despite various proposed gravity compensation techniques. As a result, there is always a measurable degree of difference between on-ground and in-orbit deployment, and this difference should be correctly understood. Taking a 12.25 m mesh antenna as an example, this chapter first gives the constant-force offloading of the ring truss by choosing a proper constant-force value, the work done by gravity on the truss can theoretically be completely balanced by the designed suspension system. Then this chapter revealed the effect of the webs' weight on the deployment of the mesh antenna, if the webs are not offloaded at all. Compared with the in-orbit result, a larger driving force is required for the on-ground test to unfold the mesh antenna during the early deployment period. This is mainly caused by the increase in the gravitational energy of the webs, considering that the mass center of the webs goes up quickly during the early deployment. With regard to the bending moments of the ring truss, the maximum bending moments of truss members for on-ground versus in-orbit deployment do show some differences, but they are not remarkable.

Two kinds of web suspension systems intended to compensate for the weight of the webs were designed and evaluated, including a constant-length-design and a constant-force-design. The difference between the two designs lies in whether a single suspension string ensures a constant length or a constant force upwards. The simulation results show that too few suspension nodes cannot restrain the height variation of the mass center of the webs, while too many suspension nodes may deform the webs. Therefore, regardless of which design is selected, the number of suspension nodes on the webs should be selected carefully, and a more moderate number seems to be a better choice.

The results described herein can directly guide the design of the gravity compensation systems for mesh antennas. Moreover, the modeling and energetic analysis methods developed here can also be extended to the analysis of other flexible mechanical systems that require gravity compensation, such as those used in robotics and aviation engineering.

4.7 Conclusions and Prospects

4.7.1 Conclusions

It is difficult for on-ground experiments to predict the in-orbit deployment behavior and the surface accuracy after deployment due to the gravitational difference between on-ground and in-orbit environment. On account of this, this thesis used simulation to optimize the designs of the deployable truss, cable networks, as well as gravity compensation systems, to minimize the difference between the in-orbit and on-ground behavior of mesh antennas, and to achieve an accurate in-orbit prediction. This thesis addresses the following three aspects:

1. The multiscale characteristics of cable-pulley systems are bottleneck restricting the dynamics simulation of antenna. This thesis proposed a multiscale dynamic method for modeling cable-pulley system, in which the cable is separated into two segments by the dynamic contact border: the non-contact segment was modeled with variant-length cable elements based on Arbitrary Lagrangian-Eulerian (ALE) formulation, and the contact segment was merely treated as a virtual segment to dynamically locate the contact border. This method avoids the problems of dense mesh and small step caused by traditional contact modeling methods, greatly improves the computational efficiency, and makes it possible to simulate antenna deployment with multiscale nature of time and length. Furthermore, two kinds of synchronization mechanisms with variant-wrap-angle and constant-wrap-angle were designed to synchronize each bay of the deployable trusses. The proposed multiscale method was used to evaluate the effects of synchronization mechanisms on the deployment dynamics of mesh antennas. Simulation results show that the driving force by motors and the bending moment of the ring truss depend on the synchronization of deployment. Variant-wrap-angle design leads to remarkable asynchronous deployment, while the constant-wrap-angle design contributes to nearly synchronous deployment, accompanied by smaller bending moments in the ring truss and smaller driving forces by motors. Further sensitivity analyses revealed that the driving force is more sensitive to the transmission efficiency than the translational friction. However, the two kinds of frictions have comparable effects on the bending moment.
2. A robust design of cable networks against variation in tension forces was proposed which reduces the accuracy requirement of the tension forces to guarantee

the tautness of the cable network and realizes the shape consistency of the cable network between on-ground and in-orbit environment. The method consists of two parts: (1) A force tolerance was defined to quantitatively characterize the robustness of the cable network to tension forces; (2) An optimization algorithm was established to enlarge the force tolerance by optimizing the tension forces and the geometry of the cable network. The proposed methodology was applied to design mesh for a 6-bay front-fed antenna and a 30-bay offset antenna. The method of robust design presented is simple and efficient, and it can be readily applied to actual scenarios.

3. The effects of on-ground gravity compensation systems for webs on deployment dynamics of mesh antennas were revealed. Compared with the in-orbit result, a larger driving force is required for the on-ground test to unfold the mesh antenna during the early deployment period. This is mainly caused by the increase in the gravitational energy of the webs, considering that the mass center of the webs goes up quickly during the early deployment. In terms of the constant-length offloading and constant-force offloading, it proves that too few or too many suspension nodes will add more constraints to the deployment, and a moderate number of suspension nodes with nearly equal distribution on webs is a better choice to reduce the difference between on-ground and in-orbit deployment.

The innovation points of this thesis are summarized as follows:

1. A multiscale dynamic method for modeling cable-pulley systems was proposed which avoids the problems of dense mesh and small step and makes it possible to simulate antenna deployment with multiscale nature of time and length. Two kinds of synchronization mechanisms with variant-wrap-angle and constant-wrap-angle were designed to synchronize each bay of the deployable trusses. The proposed multiscale method was used to evaluate the effects of synchronization mechanisms on the deployment dynamics of mesh antennas. Simulation results show that the constant-wrap-angle design leads to a nearly synchronous deployment accompanied by remarkably small driving forces.
2. A design methodology for achieving a robust cable network against variation in tension forces was proposed, which consists of two parts: first, a force tolerance was defined to quantitatively characterize the robustness of the cable network to tension forces; second, an optimization algorithm was established to enlarge the force tolerance by optimizing the tension forces and the geometry of the cable network. The proposed methodology was applied to design mesh for a 6-bay front-fed antenna and a 30-bay offset antenna. Results show that after optimization, the force tolerances were enlarged up to about four and two times respectively. This design reduces the accuracy requirement of the tension forces to ensure the tautness of the cable network.
3. The effects of on-ground gravity compensation systems for webs on deployment dynamics of mesh antennas were revealed. If the webs are not offloaded, the driving force on ground is quite different from that in orbit. When the webs are offloaded, regardless of which design is selected, the number of suspension nodes on the webs should be selected carefully, and a more moderate number seems to be a better choice.

4.7.2 Prospects

1. In view of the multiscale problem faced by the modeling of the cable-pulley system, asynchronous variational integrators with fast and slow variables [132] can be adopted at the numerical algorithm level in which each unit uses an independent time step, so as to avoid that the computational efficiency of the whole system being burdened by a certain part. The asynchronous variational integrators are implanted into the numerical algorithm of multibody systems, which has a broad development prospect.
2. Smart materials are combined with deployable mesh antennas to conduct adjustment in orbit. For example, the tension spring ties can be replaced by intelligent actuators to realize the active control of the forces of the tie forces in orbit. Shape memory polymer [133] and other materials can be used to enhance the adaptability of the antenna for in-orbit environment.
3. The designs of the deployable truss, optimization of cable networks, and simulation of deployment dynamics with gravity compensation systems can be combined with experiment to conduct validation in order to guide engineering projects.

Chapter 5
Conclusions and Prospects

5.1 Conclusions

It is difficult for on-ground experiments to predict the in-orbit deployment behavior and the surface accuracy after deployment due to the gravitational difference between on-ground and in-orbit environment. On account of this, this thesis used simulation to optimize the designs of the deployable truss, cable networks, as well as gravity compensation systems, to minimize the difference between the in-orbit and on-ground behavior of mesh antennas, and to achieve an accurate in-orbit prediction. This thesis addresses the following three aspects:

1. The multiscale characteristics of cable-pulley systems are bottleneck restricting the dynamics simulation of antenna. This thesis proposed a multiscale dynamic method for modeling cable-pulley system, in which the cable is separated into two segments by the dynamic contact border: the non-contact segment was modeled with variant-length cable elements based on Arbitrary Lagrangian-Eulerian (ALE) formulation, and the contact segment was merely treated as a virtual segment to dynamically locate the contact border. This method avoids the problems of dense mesh and small step caused by traditional contact modeling methods, greatly improves the computational efficiency, and makes it possible to simulate antenna deployment with multiscale nature of time and length. Furthermore, two kinds of synchronization mechanisms with variant-wrap-angle and constant-wrap-angle were designed to synchronize each bay of the deployable trusses. The proposed multiscale method was used to evaluate the effects of synchronization mechanisms on the deployment dynamics of mesh antennas. Simulation results show that the driving force by motors and the bending moment of the ring truss depend on the synchronization of deployment. Variant-wrap-angle design leads to remarkable asynchronous deployment, while the constant-wrap-angle design contributes to nearly synchronous deployment, accompanied by smaller

© The Author(s), under exclusive license to Springer Nature Singapore Pte Ltd. 2026

K. Fu, *Design and Prediction of In-orbit Dynamic Behavior of Deployable Mesh Antennas*, Springer Theses, https://doi.org/10.1007/978-981-92-0140-2_5

bending moments in the ring truss and smaller driving forces by motors. Further sensitivity analyses revealed that the driving force is more sensitive to the transmission efficiency than the translational friction. However, the two kinds of frictions have comparable effects on the bending moment.

2. A robust design of cable networks against variation in tension forces was proposed which reduces the accuracy requirement of the tension forces to guarantee the tautness of the cable network and realizes the shape consistency of the cable network between on-ground and in-orbit environment. The method consists of two parts: (1) A force tolerance was defined to quantitatively characterize the robustness of the cable network to tension forces; (2) An optimization algorithm was established to enlarge the force tolerance by optimizing the tension forces and the geometry of the cable network. The proposed methodology was applied to design mesh for a 6-bay front-fed antenna and a 30-bay offset antenna. The method of robust design presented is simple and efficient, and it can be readily applied to actual scenarios.
3. The effects of on-ground gravity compensation systems for webs on deployment dynamics of mesh antennas were revealed. Compared with the in-orbit result, a larger driving force is required for the on-ground test to unfold the mesh antenna during the early deployment period. This is mainly caused by the increase in the gravitational energy of the webs, considering that the mass center of the webs goes up quickly during the early deployment. In terms of the constant-length offloading and constant-force offloading, it proves that too few or too many suspension nodes will add more constraints to the deployment, and a moderate number of suspension nodes with nearly equal distribution on webs is a better choice to reduce the difference between on-ground and in-orbit deployment.

The innovation points of this thesis are summarized as follows:

1. A multiscale dynamic method for modeling cable-pulley systems was proposed which avoids the problems of dense mesh and small step and makes it possible to simulate antenna deployment with multiscale nature of time and length. Two kinds of synchronization mechanisms with variant-wrap-angle and constant-wrap-angle were designed to synchronize each bay of the deployable trusses. The proposed multiscale method was used to evaluate the effects of synchronization mechanisms on the deployment dynamics of mesh antennas. Simulation results show that the constant-wrap-angle design leads to a nearly synchronous deployment accompanied by remarkably small driving forces.
2. A design methodology for achieving a robust cable network against variation in tension forces was proposed, which consists of two parts: first, a force tolerance was defined to quantitatively characterize the robustness of the cable network to tension forces; second, an optimization algorithm was established to enlarge the force tolerance by optimizing the tension forces and the geometry of the cable network. The proposed methodology was applied to design mesh for a 6-bay front-fed antenna and a 30-bay offset antenna. Results show that after optimization, the force tolerances were enlarged up to about four and two times

respectively. This design reduces the accuracy requirement of the tension forces to ensure the tautness of the cable network
3. The effects of on-ground gravity compensation systems for webs on deployment dynamics of mesh antennas were revealed. If the webs are not offloaded, the driving force on ground is quite different from that in orbit. When the webs are offloaded, regardless of which design is selected, the number of suspension nodes on the webs should be selected carefully, and a more moderate number seems to be a better choice.

5.2 Prospects

1. In view of the multiscale problem faced by the modeling of the cable-pulley system, asynchronous variational integrators with fast and slow variables [132] can be adopted at the numerical algorithm level in which each unit uses an independent time step, so as to avoid that the computational efficiency of the whole system being burdened by a certain part. The asynchronous variational integrators are implanted into the numerical algorithm of multibody systems, which has a broad development prospect.
2. Smart materials are combined with deployable mesh antennas to conduct adjustment in orbit. For example, the tension spring ties can be replaced by intelligent actuators to realize the active control of the forces of the tie forces in orbit. Shape memory polymer [133] and other materials can be used to enhance the adaptability of the antenna for in-orbit environment.
3. The designs of the deployable truss, optimization of cable networks, and simulation of deployment dynamics with gravity compensation systems can be combined with experiment to conduct validation in order to guide engineering projects.

References

1. Maral G, Bousquet M, Sun Z. Satellite communications systems: systems, techniques and technology[M]. 5th ed. Chichester: John Wiley & Sons; 2009.
2. Ruze J. Antenna tolerance theory-a review[J]. Proc IEEE. 1966;54:633–42.
3. Wang SJ, Cameron JM. Dynamics and control of a large space antenna[J]. J Guid Control Dyn. 1984;7:69–76.
4. Meguro A, Harada S, Watanabe M. An ultra-light large antenna reflector for communication satellites[C]. In: 54th International Astronautical Congress of the International Astronautical Federation, the International Academy of Astronautics, and the International Institute of Space Law, Bremen, Germany: IAC-03-I.1.06; 2003.
5. Banik J. Realizing large structures in space. In: Frontiers of Engineering: Reports on leading-edge Engineering from the 2015 Symposium[M]. Washington DC: National Academies Press; 2016. p. 55–62.
6. Chodimella S, Moore J, Otto J, et al. Design evaluation of a large aperture deployable antenna[C]. In: 47th AIAA/ASME/ASCE/AHS/ASC Structures, Structural Dynamics, and Materials Conference, Newport, Rhode Island: AIAA-2006-1603; 2006.
7. Duan B. The state-of-the-art and development trend of large space-borne deployable antenna[J]. Electro-Mech Eng. 2017;33:1–14. (in Chinese with English abstract)
8. Guest SD, Pellegrino S. A new concept for solid surface deployable antennas[J]. Acta Astronaut. 1996;38:103–13.
9. Lichodziejewski D, Cravey R, Hopkins G. Inflatably deployed membrane waveguide array antenna for space [C]. In: 44th AIAA/ASME/ASCE/AHS Structures, Structural Dynamics, and Materials Conference, Norfolk, Virginia: AIAA 2003-1649; 2003.
10. Huang J. The development of inflatable array antennas[J]. IEEE Antennas Propag Mag. 2001;43:44–50.
11. Li T, Zhang Y, Duan B. Deployment kinematic analysis and control of hoop truss deployable antenna[J]. J Xidian Univ. 2007;34:916–21. (in Chinese with English abstract)
12. Chang H. Research on form-finding and deployment dynamics for modular cable-truss antenna[J]. J Deep Space Explor. 2017;4:325–32. (in Chinese with English abstract)
13. Zolesi VS, Ganga PL, Scolamiero L, et al. On an innovative deployment concept for large space structures[C]. In: In 42nd International Conference on Environmental Systems (ICES), San Diego, USA: AIAA 2012-3601; 2012.
14. Hasanzade V, Sedighy SH, Shahravi M. Compact deployable umbrella antenna design with optimum communication properties[J]. J Spacecr Rocket. 2017;54:782–8.
15. Scialino L, Ihle A, Migliorelli M, et al. Large deployable reflectors for telecom and earth observation applications[J]. CEAS Space J. 2013;5:125–46.

© Tsinghua University Press 2026

K. Fu, *Design and Prediction of In-orbit Dynamic Behavior of Deployable Mesh Antennas*, Springer Theses, https://doi.org/10.1007/978-981-92-0140-2

16. Feng T, Ji Y, Xiao Y, et al. Overview of space-borne perimeter truss antenna and its application[J]. Space Electr Technol. 2015;2:22–8. (in Chinese with English abstract)
17. Murphey T. Historical perspectives on the development of deployable reflectors[C]. In: 50th AIAA/ASME/ASCE/AHS/ASC Structures, Structural Dynamics, and Materials Conference, Palm Springs, California, USA: AIAA 2009-2605; 2009.
18. Takano T, Sugawara A, Sasaki S. System considerations of onboard antennas for SSPS[J]. URSI Radio Science Bulletin. 2004;2004:16–20.
19. Thomson MW. Astromesh deployable reflectors for Ku- and Ka-band commercial satellites[C]. In: 20th AIAA International Communication Satellite Systems Conference and Exhibit, Didian: AIAA-2002-2032; 2002.
20. Tibert AG, Pellegrino S. Deployable tensegrity reflectors for small satellites[J]. J Spacecr Rocket. 2002;39:701–9.
21. Li B, Qi X, Huang H, et al. Modeling and analysis of deployment dynamics for a novel ring mechanism[J]. Acta Astronaut. 2016;120:59–74.
22. Nie R, He B, Zhang L. Deployment dynamics modeling and analysis for mesh reflector antennas considering the motion feasibility[J]. Nonlinear Dyn. 2018;91:549–64.
23. Yang R. On form finding and precision adjustment technology for reflector of cable-net deployable antenna [D]. Xi'an: Xidian University; 2010. (in Chinese with English abstract)
24. Thomson MW, Marks GW, Hedgepeth JM. Light-weight reflector for concentrating radiation[P]: United States, 5680145. 1997-10-21.
25. Gilger LD, Thomson MW, Gereau JC. Micro fold reflector [P]: United States, US 6323827 B1. 2001-11-27.
26. Yang C, Ren G. Dynamic simulation of multifold deployable rings[J]. AIAA J. 2014;52:1555–9.
27. Zhang Y, Li N, Yang G, et al. Dynamic analysis of the deployment for mesh reflector deployable antennas with the cable-net structure[J]. Acta Astronaut. 2017;131:182–9.
28. Hu H, Tian Q, Zhang W, et al. Nonlinear dynamics and control of large deployable space structures composed of trusses and meshes [J]. Adv Mech. 2013;43:390–414. (in Chinese with English abstract)
29. Hu H, Tian Q, Liu C. Computational dynamics of soft machines[J]. Acta Mech Sinica. 2017:516–28.
30. Du X, Du J, Bao H, et al. Deployment analysis of deployable antennas considering cable net and truss flexibility[J]. Aerosp Sci Technol. 2018;82-83:557–65.
31. Peng Y, Zhao Z, Zhou M, et al. Flexible multibody model and the dynamics of the deployment of mesh antennas[J]. J Guid Control Dyn. 2017;40:1499–510.
32. Datashvili L, Endler S, Wei B, et al. Study of mechanical architectures of large deployable space antenna apertures: from design to tests[J]. CEAS Space J. 2013;5:169–84.
33. Agrawal PK, Anderson M, Card MF. Preliminary design of large reflectors with flat facets[J]. IEEE Trans Antennas Propag. 1981;29:688–94.
34. Meguro A, Harada S, Watanabe M. Key technologies for high-accuracy large mesh antenna reflectors[J]. Acta Astronaut. 2003;53:899–908.
35. Hedgepeth JM. Accuracy potentials for large space antenna reflectors with passive structure[J]. J Spacecr Rocket. 1982;19:211–7.
36. Miyasaka A, Hommat M, Tsujigata A, et al. Design and ground verification of large deployable reflector[C]. In: 42th MAA/ASME/ASCE/AHS/ASC Structures, Structural Dynamics, and Materials Conference and Exhibit Seattle, WA, United states: AIAA-2001-1480; 2001.
37. Peng Y. Flexible multibody model and dynamics of deployment of large mesh antennas[D]. Beijing: Tsinghua University; 2016. (in Chinese with English abstract)
38. Li P. Deployment dynamics of the large-scale hoop truss antenna of satellite[D]. Beijing: Beijing Institute of Technology; 2016. (in Chinese with English abstract)
39. Liu L, Shan J, Tao C. Simulation of non-synchronous deployment of the large deployable hoop truss antenna[C]. In: The 2nd AIAA Spacecraft Structures Conference, Kissimmee, Florida; 2015.

40. Liu L, Shan J, Zhang Y. Dynamics modeling and analysis of spacecraft with large deployable hoop-truss antenna[J]. J Spacecr Rocket. 2016;53:471–9.
41. Yang J, Xiao Y, Zhao Z, et al. Distributed driving deployment method and numerical analysis of large hoop truss antenna[J]. Sci Sin Phys Mech Astron. 2017;47:104604. (in Chinese with English abstract)
42. Bauchau OA. Flexible multibody dynamics[M]. The Netherlands: Springer Science & Business Media; 2010.
43. Geradin M, Cardona A. Flexible multibody dynamics: a finite element approach[M]. New York, NY: John Wiley; 2001.
44. Mitsugi J, Ando K, Senbokuya Y, et al. Deployment analysis of large space antenna using flexible multibody dynamics simulation[J]. Acta Astronaut. 2000;47:19–26.
45. Liu Y. Advanced dynamics[M]. Beijing: Higher education press; 2001. (in Chinese)
46. Shabana AA. Definition of the slopes and the finite element absolute nodal coordinate formulation[J]. Multibody Syst Dyn. 1997;1:339–48.
47. Gerstmayr J, Shabana AA. Analysis of thin beams and cables using the absolute nodal coordinate formulation[J]. Nonlinear Dyn. 2006;45:109–30.
48. Simo JC, Vu-Quoc L. A three-dimensional finite-strain rod model. Part II: computational aspects[J]. Comput Methods Appl Mech Eng. 1986;58:79–116.
49. Simo JC. A finite strain beam formulation. The three-dimensional dynamic problem. Part I[J]. Comput Methods Appl Mech Eng. 1985;49:55–70.
50. Tang J, Ren G, Zhu W, et al. Dynamics of variable-length tethers with application to tethered satellite deployment[J]. Commun Nonlinear Sci Numer Simul. 2011;16:3411–24.
51. Yu L, Zhao Z, Tang J, et al. Integration of absolute nodal elements into multibody system[J]. Nonlinear Dyn. 2010;62:931–43.
52. Tian Q. Flexible multibody dynamics research and application based on the absolute nodal coordinate method[D]. Wuhan: Huazhong University of Science & Technology; 2009. (in Chinese with English abstract)
53. Hong D, Ren G. A modeling of sliding joint on one-dimensional flexible medium[J]. Multibody Syst Dyn. 2011;26:91–106.
54. Zhao Z. Post-doctoral station report: modeling of flexible multibody systems and dynamics research of deployable mechanism[R]. Beijing: Tsinghua University; 2013. (in Chinese)
55. Zhao M, Guan F. Deployment dynamics analysis of circular truss deployable antenna with friction[J]. Chin J Space Sci. 2006;26:220–6. (in Chinese with English abstract)
56. Zhao M, Wu K, Guan F. Dynamic analysis of deployable space truss structures[J]. J Zhejiang Univ (Eng Sci). 2005;39:1669–74. (in Chinese with English abstract)
57. Qi Z, Chang J, Wang G. Analysis and control of deployment process for hoop truss deployable antenna[J]. J Astron. 2014;35:61–8. (in Chinese with English abstract)
58. Li T. Deployment analysis and control of deployable space antenna[J]. Aerosp Sci Technol. 2012;18:42–7.
59. Peng Y, Wei Y, Zhou M. Efficient modeling of cable-pulley system with friction based on arbitrary-lagrangian-eulerian approach[J]. Appl Math Mech. 2017;38:1785–802.
60. John R. The effect of aperture errors on the antenna radiation pattern[J]. Il Nuovo Cimento (1943–1954). 1952;9:364–80.
61. Calladine CR. Buckminster fuller's "tensegrity" structures and clerk Maxwell's rules for the construction of stiff frames[J]. Int J Solids Struct. 1978;14:161–72.
62. Pellegrino S, Calladine CR. Matrix analysis of statically and kinematically indeterminate frameworks[J]. Int J Solids Struct. 1986;22:409–28.
63. Tibert G. Optimal design of tension truss antennas[C]. In: 44th AIAA/ASME/ASCE/AHS Structures, Structural Dynamics, and Materials Conference, Norfolk, Virginia: AIAA 2003-1629; 2003.
64. Tibert G, Pellegrino S. Review of form-finding methods for tensegrity structures[J]. Int J Space Struct. 2011;26:241–55.

65. Schek H-J. The force density method for form finding and computation of general networks[J]. Comput Methods Appl Mech Eng. 1974;3:115–34.
66. Linkwitz K. Formfinding by the "direct approach" and pertinent strategies for the conceptual design of prestressed and hanging structures[J]. Int J Space Struct. 1999;14:73–87.
67. Tibert G. Deployable tensegrity structures for space applications[D]. Stockholm: Royal Institute of Technology; 2002.
68. Morterolle S, Maurin B, Quirant J, et al. Numerical form-finding of geotensoid tension truss for mesh reflector[J]. Acta Astronaut. 2012;76:154–63.
69. Di J. On optimal adjusting technology for reflector precision of the cable-net deployable antenna structures[D]. Xi'an: Xidian University; 2005. (in Chinese with English abstract)
70. Yang D. Structure design and profile adjustment of large deployable mesh antenna for satellite[D]. Xi'an: Xidian University; 2010. (in Chinese with English abstract)
71. Du J, Zong Y, Bao H. Shape adjustment of cable mesh antennas using sequential quadratic programming[J]. Aerosp Sci Technol. 2013;30:26–32.
72. Li P, Liu C, Tian Q, et al. Dynamics of a deployable mesh reflector of satellite antenna: form-finding and modal analysis[J]. J Comput Nonlinear Dyn. 2016;11:04101701–12.
73. Belvin WK, Herstrom CL, Edighoffer HH. Quasistatic shape adjustment of a 15-meter-diameter space antenna[J]. J Spacecr Rocket. 1989;26:129–36.
74. Zong Y, Hu N, Duan B, et al. Manufacturing error sensitivity analysis and optimal design method of cable-network antenna structures[J]. Acta Astronaut. 2016;120:182–91.
75. Du J, Wang C, Bao H, et al. Robust shape adjustment with finite element model updating for mesh reflectors[J]. AIAA J. 2017;55:1450–9.
76. Yang D, Du J, Li T. Optimization method for shape adjustment of perimeter-truss mesh antenna reflectors[J]. China Mech Eng. 2010;21:35–7. (in Chinese with English abstract)
77. Tanaka H. Surface error estimation and correction of a space antenna based on antenna gainanalyses[J]. Acta Astronaut. 2011;68:1062–9.
78. Zhang S, Du J, Duan B, et al. Integrated structural—electromagnetic shape control of cable mesh reflector antennas[J]. AIAA J. 2014;53:1395–9.
79. Zhang S, Du J, Duan B. Integrated structural electromagnetic analysis of mesh reflectors with structural random dimensional errors[J]. AIAA J. 2015;53:2838–44.
80. Tolson RH, Huang J-K. Integrated control of thermally distorted large space antennas[J]. J Guid Control Dyn. 1992;15:605–14.
81. Hu N, Bao H, Duan B, et al. Topology optimization of reflector antennas based on integrated thermal-structural-electromagnetic analysis[J]. Struct Multidiscip Optim. 2016;55:715–22.
82. Tabata M, Natori MC. Active shape control of a deployable space antenna reflector[J]. J Intell Mater Syst Struct. 1996;7:235–40.
83. Tang Y, Li T, Ma X. Pillow distortion analysis for a space mesh reflector antenna[J]. AIAA J. 2017;55:3206–13.
84. Zhang S, Yang G, Zhang Y. An approximation mathematical formula of pattern analysis for distorted reflector antennas considering surface normal vector variation[J]. Int J Antenn Propag. 2016;2016:1–6.
85. Mitsugi J, Yasaka T, Miura K. Shape control of the tension truss antenna[J]. AIAA J. 1990;28:316–22.
86. Meguro A, Ishikawa H, Tsujihata A. Study on ground verification for large deployable modular structures[J]. J Spacecr Rocket. 2006;43:780–7.
87. Meguro A, Mitsugi J. Ground verification of deployment dynamics of large deployable space structures[J]. J Spacecr Rocket. 1992;29:835–41.
88. Mao H, Ganga PL, Ghiozzi M, et al. Deployment of bistable self-deployable tape spring booms using a gravity offloading system[J]. J Aerosp Eng. 2017;30:04017007.
89. Qin G. Research on underwater photogrammetry for surface measurement of satellite antenna in simulated zero-gravity conditions[D]. Zhengzhou: Information Engineering University of the People's Liberation Army; 2011. (in Chinese with English abstract)

90. Peng H, He B. A new gravity compensation method of space-borne perimeter truss deployable reflectors[J]. China Mech Eng. 2019;30:379–84. (in Chinese with English abstract)
91. Xu W, Liang B, Li C, et al. A review on simulated micro-gravity experiment systems of space robot[J]. Robot. 2009;31:88–96. (in Chinese with English abstract)
92. Russo A, Sinatra R, Xi F. Static balancing of parallel robots[J]. Mech Mach Theory. 2005;40:191–202.
93. Streit DA, Shin E. Equilibrators for planar linkages[J]. J Mech Des. 1993;115:604–11.
94. Ulrich N, Kumar V. Passive mechanical gravity compensation for robot manipulators[C]. In: 1991 IEEE International Conference on Robotics and Automation, Sacramento, California; 1991. p. 1536–41.
95. Wang J, Gosselin CM. Static balancing of spatial three degree freedom parallel mechanism[J]. Mech Mach Theory. 1999;34:437–52.
96. Wang J, Gosselin CM. Static balancing of spatial four-degree-of freedom parallel mechanisms[J]. Mech Mach Theory. 2000;35:563–92.
97. Li Y, Shao Z, Tian S, et al. Analysis and evaluation on unloading ratio of zero-g simulation device of space manipulator based on suspension system[J]. Robot. 2016;38:293–300. (in Chinese with English abstract)
98. Fischer A, Pellegrino S. Interaction between gravity compensation suspension system and deployable structure[J]. J Spacecr Rocket. 2000;37:93–9.
99. Greschik G, Belvin WK. High-fidelity gravity offloading system for free-free vibration testing[J]. J Spacecr Rocket. 2007;44:132–42.
100. Wu M, Zhang T, Xiang P, et al. Single-layer deployable truss structure driven by elastic components[J]. J Aerosp Eng. 2019;32:04018144.
101. Cheng L, Guan F, Zhang H. Deployment experiment of deployable truss antenna[J]. Chin J Space Sci. 2010;30:79–84. (in Chinese with English abstract)
102. Sugimoto T, Tsunoda H, Hariu K-I, et al. Structural design and deployment test methods for a large deployable mesh reflector[C]. In: 38th Structures, Structural Dynamics, and Materials Conference, Kissimmee, FL: AIAA-97-1148; 1997.
103. Tsunoda H, Hariu K-I, Kawakami Y, et al. Deployment test methods for a large deployable mesh reflector[J]. J Spacecr Rocket. 1997;34:811–6.
104. Takano T, Natori M, Miyoshi K, et al. Characteristics verification of a deployable onboard antenna of 10 m maximum diameter[J]. Acta Astronaut. 2002;51:771–8.
105. Xu H. Deployment dynamics, control and experiment of deployable antennas[D]. Xi'an: Xidian University; 2014. (in Chinese with English abstract)
106. Li P, Qin MA, Song Y, et al. Deployment dynamics simulation and ground test of a large space hoop truss antenna reflector[J]. Sci Sin Phys Mech Astron. 2017;47:104602. (in Chinese with English abstract)
107. Fu K, Zhao Z, Ren G, et al. From multiscale modeling to design of synchronization mechanisms in mesh antennas[J]. Acta Astronaut. 2019;159:156–65.
108. Sauer T. Numerical analysis[M]. Boston, MA: Pearson; 2012.
109. Li J, Zhang X. Theoretical mechanics[M]. 2nd ed. Beijing: Tsinghua University; 2010. (in Chinese with English abstract)
110. Zhao Y, Shi P. Modeling theory and method of spacecraft flight dynamics[M]. Beijing: Beihang University Press; 2012. (in Chinese with English abstract)
111. Li M, Wang R, Zhao Z. Passive nonlinear actuators for deploying mesh antennas[J]. Acta Astronaut. 2018;152:254–61.
112. Fang K, Sun X. Shape preserving interpolation by quintic parametric curve[J]. J Numer Methods Computer. 2002;1:24–30. (in Chinese with English abstract)
113. Liu Z, Xiong M, Ye Y. Mechanical competitions and modeling for college students[M]. Beijing: China Water&Power Press; 2012. (in Chinese)
114. Sun F, Li Q. Research on involute linetype of cam for buoyancy compensation of bell prover[J]. J Astron Metrol Measur. 2013;33:42–5. (in Chinese with English abstract)

115. Zhao Z, Ren G. A quaternion-based formulation of Euler-Bernoulli beam without singularity[J]. Nonlinear Dyn. 2012;67:1825–35.
116. Wu J, Zhao Z, Ren G. Multibody analysis of the force in deploying booms[J]. J Guid Control Dyn. 2013;36:1881–6.
117. Sun FJ, Zhu ZH, Larosa M. Dynamic modeling of cable towed body using nodal position finite element method[J]. Ocean Eng. 2011;38:529–40.
118. Cha H-Y, Choi J, Ryu HS, et al. Stick-slip algorithm in a tangential contact force model for multi-body system dynamics[J]. J Mech Sci Technol. 2011;25:1687–94.
119. Fu K, Du J, Li J, et al. Robust design of tension truss antennas against variation in tension forces[J]. AIAA J. 2018;55:1450–9.
120. Pellegrino S. Structural computations with the singular value decomposition of the equilibrium matrix[J]. Int J Solids Struct. 1993;30:3025–35.
121. Du J. Structural optimization and its application to vibration and acoustic design[M]. Beijing: Tsinghua University Press; 2015. (in Chinese)
122. Bendsoe MP, Olhoff N, Taylor JE. A variational formulation for multicriteria structural optimization[J]. J Struct Mech. 1983;11:523–44.
123. Chen B. Optimization theory and algorithms [M]. 2nd ed. Beijing: Tsinghua University Press; 2005. (in Chinese)
124. Rathod HT, Nagaraja KV, Venkatesudu B, et al. Gauss legendre quadrature over a triangle[J]. J Indian Inst Sci. 2004;84:183–8.
125. Yang D, Liu J, Zhang Y, et al. Optimal surface profile design of deployable mesh reflectors via a force density strategy[J]. Acta Astronaut. 2017;130:137–46.
126. Greene WH. Effects of random member length errors on the accuracy and internal loads of truss antennas[J]. J Spacecr Rocket. 1985;22:554–9.
127. Nocedal J, Wright SJ. Numerical optimization[M]. Madison: Springer; 2006.
128. Rudge AW, Adatia NA. Offset-parabolic-reflector antennas: a review[J]. Proc IEEE. 1978;66:1592–618.
129. Angevain J-C, Rodrigues G, Santiago-Prowald J, et al. Phyllotactic arrangements of reflector mesh facets to decrease grating lobes[C]. In: In 2015 9th European Conference on Antennas and Propagation (EuCAP), Lisbon, Portugal; 2015.
130. Peng Y, Yang J, Xiao Y, et al. Gravity effect on deployment dynamics of astromesh[J]. Eng Mech. 2018;35:226–34. (in Chinese with English abstract)
131. Zhao Z, Fu K, Li M, et al. Gravity compensation system of mesh antennas for in-orbit prediction of deployment dynamics[J]. Acta Astronaut. 2020;167:1–13.
132. Lew A, Marsden JE, Ortiz M, et al. Asynchronous variational integrators[J]. Arch Ration Mech Anal. 2003;167:85–146.
133. Liu Y, Du H, Liu L, et al. Shape memory polymers and their composites in aerospace applications: a review[J]. Smart Mater Struct. 2014;23:023001.

The manufacturer's authorised representative in the EU is Springer Nature Customer Service Centre GmbH, Europaplatz 3, 69115 Heidelberg, Germany. If you have any concerns regarding our products, please contact ProductSafety@springernature.com

Printed and bound by CPI Group (UK) Ltd, Croydon, CR0 4YY
11/10/2026
02231451-0001